ELECTRONIC COMMUNICATIONS TECHNOLOGY

James K. Hardy

A Reston Book
Prentice-Hall
Englewood Cliffs, New Jersey

Library of Congress Cataloging-in-Publication Data
Hardy, James K.
  Electronic communications technology.

  "A Reston book."
  Includes index.
  1. Telecommunication. I. Title.
  TK5105.H358 1986   621.38'0413   85-25839
  ISBN 0-8359-1604-9

A Reston Book
Published by Prentice-Hall
A Division of Simon & Schuster, Inc.
Englewood Cliffs, New Jersey 07632

10 9 8 7 6 5 4 3

Printed in The United States of America

Cover photograph courtesy of the Megatek Corporation.

To Rebel
and
John, Lara, Shawna, and Sandra

CONTENTS

7. TRANSMITTERS   **227**

8. RECEIVERS   **255**

9. PHASE-LOCKED LOOPS   **281**

1. INTRODUCTION TO COMMUNICATIONS

Electronic communications is the science of sending information over electronic paths of various lengths. Along with the transmitted signals goes the hope that the information will be received and understood since no communication system is perfect. Some forms of error and distortion will always occur but they should be rare.

The information sent over these paths could be black and white or color television signals with their sound channels, stereo music, instructions for aircraft pilots or taxi drivers, telephone calls, or computer messages to name a few. The paths could be a few meters or a few kilometers or from Saturn to Earth.

In this and the following chapters we will look at the theories, circuits, and systems that make it all possible and also at the problems that prevent the communications from being perfect.

1.1 BASIC COMMUNICATION SYSTEM

In Figure 1.2 we see the three elements of any communication system: the "transmitter," the medium or path that the signals follow for whatever distance is involved, and the "receiver." The information in this case is a sound wave (variation in air pressure) so two transducers have been included to convert first from air motion into an electrical signal and then back again. The transducers are more commonly known as a microphone and a loudspeaker. Every component of this system will affect the information being carried in some way. The end result will be that the received information will never be exactly the same as the original information. This is shown in the waveforms. With careful design and a little luck, the difference between the two signals will be insignificant. But, with poor circuit design or changes in the transmission medium, the received signal could be either a poor reproduction of the original or, possibly, a meaningless mess.

1

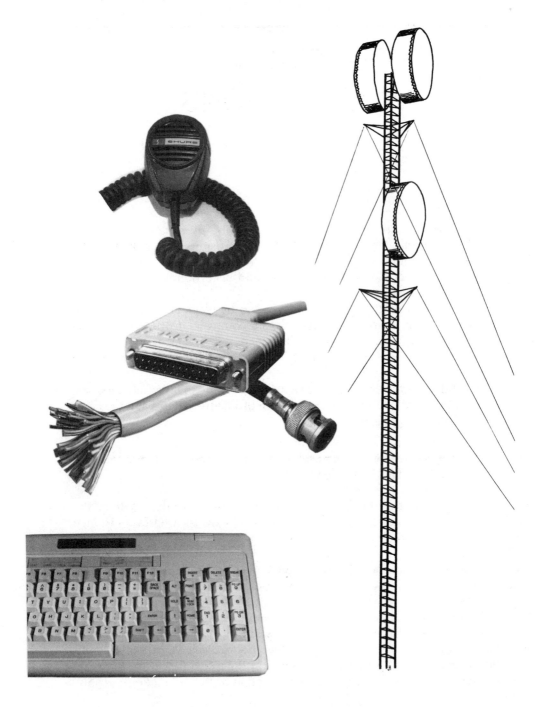

Figure 1.1 Collection of typical components from electronic communication systems.

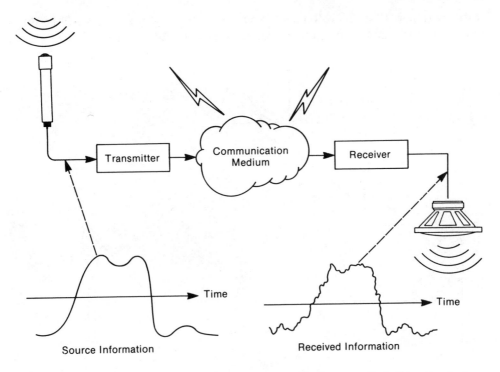

Figure 1.2 The three components of any communication system with input and output waveforms indicating that distortion has occurred.

1.1.1 The Path

Whatever the distance involved, some path or medium is needed to carry the signals between the two points. The selection can be made from a simple pair of wires, coaxial cable, waveguide, fiber optic cable, or radio waves traveling through free space, as shown in Figure 1.3. For shorter distances, wires are often used. As longer distances are encountered, any of the wire systems weaken the signals excessively so amplifiers are added periodically. This method is used, for example, in cable television systems and transatlantic cables. For longer distances and wider bandwidths, the cost of the wire and amplifiers becomes prohibitive.

It is then that the free space radio wave (wireless) becomes more attractive. Also, where mobility is required, as for aircraft, boat, and car communications, the only possibility is radio. Recently, it has been shown that fiber optic cable can provide very low-loss communication over long distances and carry a very large amount of information—a tremendous improvement over wire and cable systems. All paths will alter the signals in some way and this will be discussed in the chapters on propagation and transmission lines.

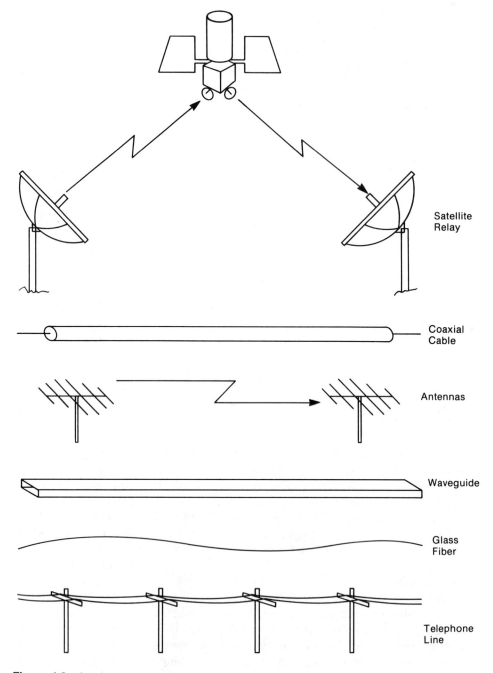

Figure 1.3 A selection of path possibilities.

1.1.2 The Transmitter

The transmitter takes the electrical signals that represent the information to be sent and prepares it for whatever medium or path is to be used. For a short-range intercom system, for example, a pair of wires will be used as the path and they will carry electrical signals at voice frequencies. In such a case, the transmitter will simply be an audio amplifier.

A transmitter for the 27 MHz Citizen's Band (CB), on the other hand, is considerably more complex. In addition to a microphone and audio amplifier, we also need a radio frequency (RF) signal capable of an average power of 6 watts at the desired channel frequency and a circuit that will change the amplitude of this "carrier" signal up and down in step with the voice signal. This transmitter is "amplitude modulating" (AM) the RF carrier. The envelope at the top and the bottom of the output signal will resemble the voice signal as shown in Figure 1.4. Other transmitters may "frequency modulate" (FM), "phase modulate" (PM), or "pulse modulate" their carriers.

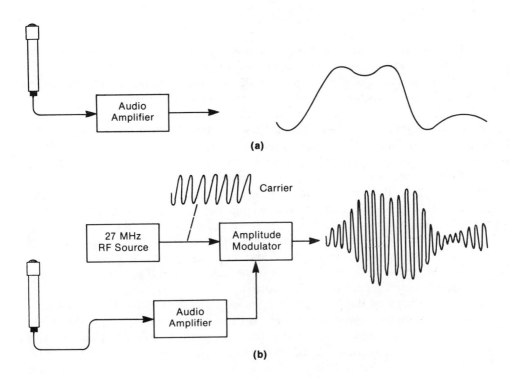

Figure 1.4 Two transmitters: (a) is simply an audio amplifier and (b) includes an amplitude modulator.

We will look at modulation in detail in Chapters 5 and 6. But why modulate? An intercom system does not use modulation. Its transmitters and receivers are connected by a pair of wires that carry electrical signals at the same frequencies as the voice signals. The system works well until someone else decides to hook an extra transmitter and receiver onto the same pair of wires. The two users will then interfere with each other because they use the same range of voice frequencies. Two CB sets would also interfere with each other if they were set to the same channel. But now, because CB uses a modulation system, one pair of talkers can switch to a new channel and a new radio carrier frequency (if one is clear) and still send voice signals back and forth. Therefore, the benefits of modulation are that many channels can share the same path or medium without interfering, just by using a different carrier frequency. Even the telephone company uses modulation on its wires and cables—not for the pair of wires coming into your house, but for the trunk lines between switching offices. In this case, the ability to place perhaps 100 phone calls on the one cable makes that cable more efficient and reduces the cost per phone call.

1.1.3 The Receiver

The last part of a communication system is the receiver. It must be chosen to match the transmission medium used. For a simple, short-range intercom system, the receiver will likely be just a loudspeaker to change the electrical signals back into acoustical signals. For a more elaborate intercom system using 250 meters of wire between transmitter and receiver, an amplifier may be required at the receiver to make up for losses in the very long lines. For a modulated system, the receiver must use the correct type of "demodulator" to recover the audio signals being sent. In this case, the receiver must also be designed to tune to the correct range of frequencies and amplify the weak signals adequately. Receivers for the AM broadcast band, for example, must tune 540 to 1600 kHz, do not need very much gain, and use a demodulator for signals that are amplitude modulated. Receivers for color television must tune several groups of frequencies between 54 and 900 MHz, require a large gain, and need three different demodulators—one for the amplitude-modulated picture, a second for the phase-modulated color information, and a third for the frequency-modulated sound, all from one TV station.

Remember that the purpose of the transmitter and receiver and the interconnecting path is to carry information with as little distortion as possible over some particular distance. Each of the three components will always distort the signal in some way so that the received signal is never the same as the original. Sometimes the difference is impossible to detect; at other times the difference is so great that the system is unusable. How great the difference must be before the communication system cannot be used is up to the individual user. Computer communications, for example, demand extremely low distortion so that

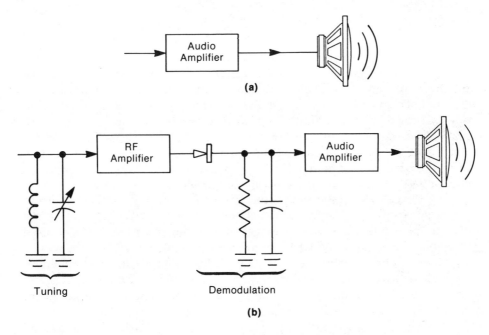

Figure 1.5 Two receivers: (a) is simply an audio amplifier to make up for path losses while (b) adds a tuned RF amplifier and demodulator.

less than one mistake occurs for every 10^9 pieces (or bits) of information sent. Radio amateurs, on the other hand, are reasonably happy even if half the words are garbled. Most other systems lie between these two extremes.

1.2 DISTORTIONS

As already mentioned, the received signal in any communication system will always be a slightly distorted version of the transmitted one. There are many things that can cause this; one is losses in the transmission path. No matter how short the path, or what type, some of the transmitted energy will be lost and loss will usually increase with distance for either wire or radio (wireless) systems. This loss of signal may seem a very minor problem since an amplifier could easily boost the weak signal back to its original size but, every amplifier adds some noise to the signal and so a distortion occurs. For long distances, the signal can become so weak that it cannot be distinguished from the noise and it is lost.

1.2.1 White Noise

Noise appears in many forms but what concerns us first is "white" noise. This is a totally random voltage or current that contains all frequencies of equal amplitude (like white light from which the name comes). If heard from a speaker it will sound like a soft hiss. The appearance of white noise voltage on an oscilloscope is shown in Figure 1.6, along with a curve that shows the relative probability of any particular instantaneous voltage occurring. The waveform shows the complete randomness of the signal that results from all the frequencies being added together and the curve shows that small amplitude signals and plenty of zero crossings are the most likely to occur.

But, the curve also shows that there is a small but finite chance that infinitely large positive or negative peak voltages could occur. White noise cannot, therefore, be measured as peak voltages and so the rms value must be used. One more point is important when we are measuring white noise. If the noise is truly white, it contains every frequency right up to infinity, with constant amplitude. Now, if we use any voltmeter, amplifier, or filter during the measurement, their bandwidths will limit the rms amplitude of the noise and so bandwidth must always be considered when noise is measured.

Now let us see where noise comes from. There are two major sources; one is every resistor and resistance (losses in other components) of any circuit and the other is the random movement of electrons through vacuum tubes, diodes, and bipolar and field effect transistors. We will look at resistor noise first.

Each resistance in a circuit, whether actual resistors, the internal resistance of signal generators, or resistive losses in other components, will generate white noise. This is caused by the random motion of electrons inside the resistor due to temperature and so is referred to as *thermal noise*. This is quite independent of any DC current passing through; even an unconnected resistor lying

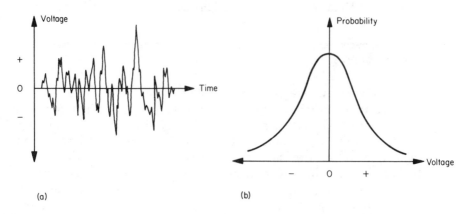

Figure 1.6 (a) Typical white-noise voltage and (b) the relative probability of any voltage occurring.

on the table will produce a noise voltage between its two leads. The rms value of this voltage is given by:

$$e_n = \sqrt{4KTBR} \text{ volts (rms)}$$

where K = Boltzmann's constant

$\quad = 1.38 \times 10^{-23}$ Joules/° Kelvin

T = absolute temperature (°C + 273)

R = resistance (Ω)

B = bandwidth (Hz)

(1.1)

Example 1.1

Back in Figure 1.2 we had a microphone producing an electrical signal. If it has an internal resistance of 15,000 ohms and sits at room temperature, how much thermal noise is being added to the voice signal? The bandwidth will be 4 kHz, which is sufficient for voice.

Room temperature will be taken as + 20°C or 293°K.

$$
\begin{aligned}
e_n &= \sqrt{4KTBR} \\
&= \sqrt{4 \times 1.38 \times 10^{-23} \times 293 \times 4000 \times 15,000} \\
&= \quad 0.985 \; \mu \text{ volts rms}
\end{aligned}
$$

This certainly doesn't look like much noise but, remember that microphones are usually connected to fairly high-gain amplifiers so that the noise voltage will be amplified right along with the voice signal voltage. The ratio of the signal voltage to the noise voltage at any point in a circuit can be used to indicate how serious the noise is. Figure 1.7 illustrates the difference between what ideally should come out of the microphone and what actually does. It also shows that the seriousness of the noise depends on the relative amplitudes of the original signal and the noise—in other words, on the signal-to-noise ratio.

The second major noise source is the so-called "shot" noise produced when current flows through diodes and transistors, etc. The noise is again white but is more difficult to calculate since each device is slightly different. The amount of noise increases with the DC bias current of the device and the bandwidth used. For a semiconductor diode,

$$i_n = \sqrt{2qIB} \text{ amperes (rms)}$$

where q = charge on an electron

$\quad = 1.60 \times 10^{-19}$ (Coulombs)

(1.2)

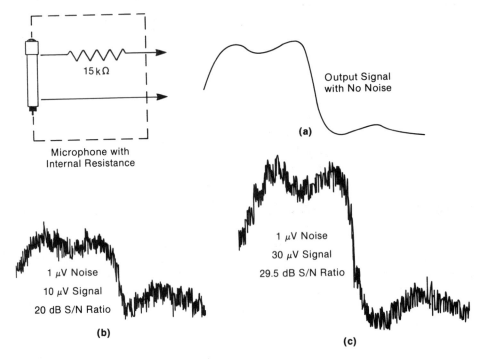

Figure 1.7 The internal resistance of the microphone adds thermal noise to the voice signal. If the voice signal is weak, the noise is serious; if it is stronger, the noise is not as noticeable.

I = bias current (Amps)
B = bandwidth (Hz)

1.2.2 Amplifier Noise Figure

The added shot noise that originates in transistors and vacuum tubes means that, even though the signals get bigger, the output signal-to-noise ratio will be worse than the input signal-to-noise ratio. Remember that the input signal to any system will already have some noise with it because the signal must come from somewhere and that source will always have some internal resistance that will generate thermal noise. A measurement of how much worse the output signal-to-noise ratio of an amplifier is than the input is given by the noise factor and the corresponding logarithmic noise figure,

$$\text{Noise factor} = \frac{S_i/N_i}{S_o/N_o} \tag{1.3}$$

$$\text{Noise figure} = 20 \log_{10} \left(\frac{S_i/N_i}{S_o/N_o} \right) \text{dB} \tag{1.4}$$

where S_i = input signal voltage
  N_i = input thermal noise voltage from the source
    resistance
  S_o = output signal voltage
    (S_i × amplifier voltage gain)
  N_o = total output noise voltage
    (amplified N_i plus added shot noise)

This is illustrated in Figure 1.8. The noise added to the signal as it passes through the amplifier comes partly from shot noise due to DC current flow and partly from thermal noise due to resistive components inside the transistor. The noise figure of a transistor will depend on the actual transistor chosen, its DC bias current, operating frequency, and relationship between source resistance and transistor input resistance.

The variation of noise figure for a range of better-quality transistors is shown in Figure 1.9. Over a wide range of frequencies, noise figures of 1 to 2 dB are possible. At higher frequencies, things get worse at the rate of 6 dB per octave (a 2:1 frequency range) mainly due to the normal loss of gain that transistors undergo at these frequencies. At very low frequencies, there is a 3 dB per

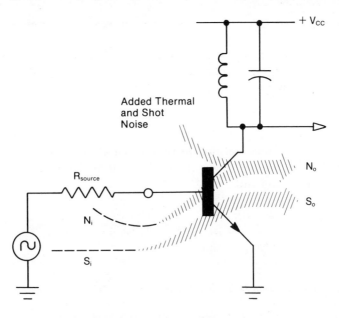

Figure 1.8 The output S/N ratio is worse than the input S/N ratio because of thermal and shot noise added inside the transistor.

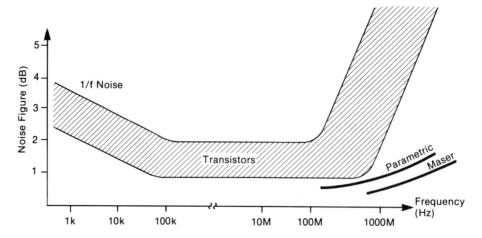

Figure 1.9 Typical noise figure variation with frequency for bipolar and field effect transistors and microwave amplifiers.

octave worsening caused by a "trapping" phenomena in the semiconductor material.

Example 1.2

Calculate the noise figure of a transistor amplifier that has an input signal of 10 microvolts[1] from a 75 ohm signal generator, a noise bandwidth of 12 kHz, a voltage gain of 25, and a total output noise of 5.0 microvolts.

Solution:

Noise figures are usually assumed to be measured at a room temperature of 20°C or 293°K. The thermal noise from the 75 ohm signal generator over the amplifier's 12 kHz bandwidth is,

$$e_n = \sqrt{4KTBR}$$
$$= \sqrt{4 \times 1.38 \times 10^{-23} \times 293 \times 12 \times 10^3 \times 75}$$
$$= 0.1206 \ \mu \text{ volts rms}$$

---

[1]The voltages used in the example for input signal and input noise from the signal generator are assumed to be "open-circuit" voltages. As soon as the generator is connected to the amplifier, the amplifier's input resistance will lower both voltages (by the same amount). Since we are using a ratio of the input signal and input noise, it doesn't really matter whether open-circuit voltages or voltages under load are used as long as they are both measured the same way.

The noise figure can then be calculated,

$$\text{Noise figure} = 20 \log_{10} \left(\frac{S_i/N_i}{S_o/N_o} \right)$$

$$= 20 \log_{10} \left(\frac{S_i/N_i}{25 \times S_i/N_o} \right)$$

$$= 20 \log_{10} \left(\frac{10/0.1206}{25 \times 10/5.0} \right)$$

$$= 4.39 \text{ dB}$$

Now, just for interest, we can find the total effective input noise to the amplifier in our example. We take the total output noise and divide by the voltage gain,

$$\frac{5.0 \, \mu \text{ V}}{25} = 0.2 \, \mu \text{ V}$$

(0.121 μ V is due to generator thermal noise and the remainder, or 0.079 μ V, is due to added transistor noise.) The ratio of total effective input noise to the source resistance noise is 4.39 dB—the same as the noise figure.

From this we can get an idea of how weak a signal we can detect with this amplifier. Figure 1.10 shows what the result of a 0.2 μ V square wave and 0.2 μ V of band-limited (12 kHz) white noise would look like. This is called the *tangential sensitivity* of an amplifier. A television picture with this signal-to-noise ratio would be unwatchable, whereas a Morse code signal could be readily understood.

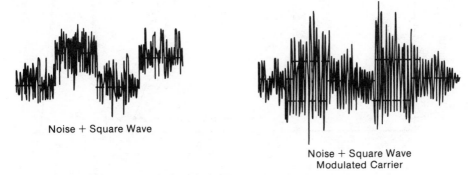

Noise + Square Wave

Noise + Square Wave
Modulated Carrier

Figure 1.10 Tangential sensitivity. Signal and noise are of equal amplitude.

The difference between the relative usefulness of the Morse code and the television signals is due to the difference in the "information" each contains. The Morse signal is either there or it isn't. It is a digital signal. Several ON and OFF sequences combine to form, for example, the letter "W" ($\cdot - -$). The amount of "information" in the waveform at any given time is, therefore, quite low. A television signal, on the other hand, is an analog signal and every change in amplitude means a change in color and brightness on the face of the picture tube. Therefore, the "information" in the TV signal is much higher and any disruption is more easily noted.

Signal-to-noise ratio, therefore, does not tell the complete story. The user has the final decision about what signal-to-noise ratio is needed for a particular application. A television picture with a signal-to-noise ratio of 25 dB would show noticeable "snow." A Morse signal with a signal-to-noise ratio of 10 dB is quite acceptable.

1.2.3 Noise Bandwidth

In Example 1.1, reference was made to the "noise bandwidth" of an amplifier; what does that mean? In Chapter 3 we will be studying filter circuits in detail but for now we can get by with a few simple points. In Figure 1.11, the frequency responses of three different filters are shown. At their midpoints, all these filters let signals through without any loss; but, at both higher and lower frequencies, the losses increase. What interests us here is how fast this loss changes with frequency. All three filters have the same bandwidth as measured between their 3 dB loss points, but, at their 20 dB loss points, for example, there is a tremendous difference. This extra bandwidth means that weak but signifi-

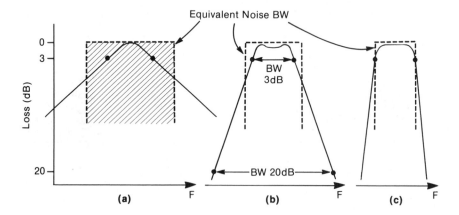

Figure 1.11 Three filter responses with identical 3 dB bandwidths but different noise bandwidths.

cant amounts of extra noise will be added just as if the bandwidth had been wider and the sides steeper. The equivalent noise bandwidth of (a) will be much wider than its 3 dB bandwidth while that of (c) will be only slightly wider.

1.2.4 External Noise

The white noise sources we have just been talking about were internal to the communication circuits. Now we look at noise that comes from outside sources. These sources might be:

1. *Man-made*—such as automobile ignition systems, fluorescent lights, electrical arcing from switchgear, and commutator-brush motors. This noise is more severe near cities and industrial areas.

2. Atmospheric—commonly referred to as *static*. This is caused by electrical discharges in the air usually associated with thunderstorms. These storms don't have to be local; even storms thousands of miles away produce signals that propagate for great distances.

3. Extraterrestrial—the sun, moon, planets, and all stars act as sources of radio signals for several reasons. First of all, any hot object radiates energy, some of which is visible if the object is hot enough. Then there are other signals from solar flares and mysterious creaks and groans from some of the planets. These noises may seem relatively insignificant but they become very disturbing to highly directional antennas tracking spacecraft, especially, for example, if one passes in front of the sun.

The relative strength of all these sources is shown in Figure 1.12. Notice that there are tremendous differences between day and night signal levels due to changes in signal propagation, and differences between rural and industrial areas.

The most important thing we can learn from Figure 1.12 is that the external noise is very strong at the lower frequencies and becomes weaker as frequency increases. This means that transmitters at low frequencies must use very high power to overcome the noise. If large receiving antennas and expensive low-noise amplifiers are used at these frequencies, the result will just be more noise picked up along with more signal. The signal-to-noise ratio will not change in spite of all the extra money and effort. This is the situation encountered in the AM broadcast band (540 to 1600 kHz). Transmitters of 50 kilowatts are common and daytime ranges of only 100 kilometers result. Receivers need very few stages since the signals are so strong, and antennas are highly inefficient ferrite loops. At greater distances, the signals are still quite strong but so is the background noise so range is limited.

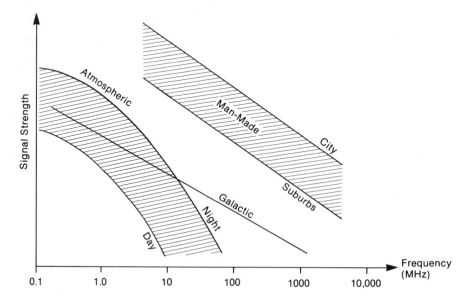

Figure 1.12 External noise levels change with frequency and location.

At higher frequencies, such as the 108 to 136 MHz aircraft band, distances of 150 kilometers with only 20 watt transmitters are possible as long as the aircraft is high enough to maintain "line of sight" conditions. At these frequencies, background noise is much lower and so high-gain, low-noise amplifiers and efficient antennas are worthwhile. The noise generated inside the receiver is now more significant than that generated outside.

1.2.5 Amplitude Distortion

Noise is not the only problem that can interfere with the quality of a communication signal. Another source of distortion is the nonlinear gain of amplifiers and some passive components (diodes, transformer cores, rusted connections, etc.). *Nonlinear* means that small signals are not amplified the same as large signals, harmonics of the applied signals are produced, and more complicated problems such as cross-modulation and intermodulation can cause unwanted signals to appear (more on this in Chapter 7).

One simple example of nonlinear gain is shown in Figure 1.13. In (a) we see that there is a curved relationship between the input and output voltages of an amplifier and that this curve can be expressed as an equation:

$$E_{out} = 10\,E_{in} + 3\,E^2_{in}$$

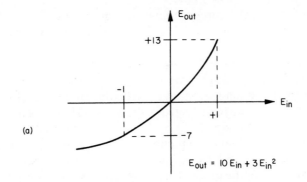

(a)

$$E_{out} = 10\,E_{in} + 3\,E_{in}^2$$

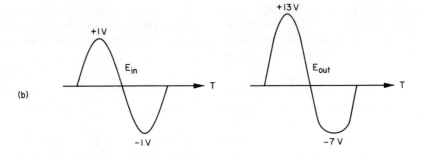

(b)

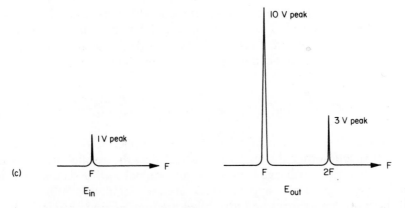

(c)

Figure 1.13   Typical transfer characteristic for an amplifier with second-order distortion (a). The resulting distortion of an input sine wave is shown in (b), and the corresponding frequency spectrums of the input and output waves are shown in (c).

In (b) of the same figure, we have an input sine wave and the corresponding output waveform showing a high peak on the positive side and a shorter, rounded peak on the negative side. The distortion is obvious. In (c) we are looking at the same input and output waveforms but this time the horizontal axis is frequency. This type of display would be seen on a "spectrum analyzer." The input spectrum shows one signal at a particular frequency (F) with a peak amplitude of 1 volt. The output spectrum shows two signals. The original signal at frequency (F) has been amplified by 10 and a new signal or harmonic has appeared at twice the original frequency and with a 3 volt peak amplitude. This type of distortion is called *second harmonic distortion* because the new signal has appeared at twice the original signal frequency.[2] The distortion is caused by the *square law gain* of this amplifier—so called because of the square term in the equation that describes the curve.

For the square law example we are dealing with, the second harmonic was 30% of the amplitude of the fundamental at the amplifier's output, but this was for a 1 volt peak input signal. For smaller input signals, the "relative size" of the second harmonic will be reduced and, for larger inputs, the harmonic's "relative size" will be increased. This relationship can be plotted on a graph as shown in Figure 1.14. The amplitudes of the fundamental output and second harmonic output can now be determined for any desired input. Notice that both scales are logarithmic.

A term often used by manufacturers of communication amplifiers to describe how linear their amplifiers are is *intercept point*. On our graph for the amplifier with second-order distortion, the "second-order intercept point" occurs at an input of 3.33 volts peak. At this point, the fundamental and second harmonic outputs will be equal in amplitude at 33.3 volts each.

We have made it sound as if nonlinear amplifiers are a bad thing; this isn't always true. Amplifiers with third- and higher-order distortion are almost always a problem, but second-order distortion in amplifiers is occasionally useful.

1.2.6 Second-Order Gain with Two Inputs

When two signals with different frequencies are applied to the input of an amplifier with second-order distortion characteristics, the results are a bit more complex. Each input will again produce an original and a second harmonic output but new signals also appear. These are the "sum" and "difference" signals of the input frequencies and are called *intermodulation products*. This distortion can occur in a Hi-Fi amplifier and produces a very harsh sound because

---

[2] The relationship between the fundamental and second harmonic is rather easily obtained just by using the 10 and 3 coefficients in the square law equation. However, for third-order distortion, a trigonometric expansion is required. For further information see *High Frequency Circuit Design* by J. Hardy, Reston Publishing Co., 1979.

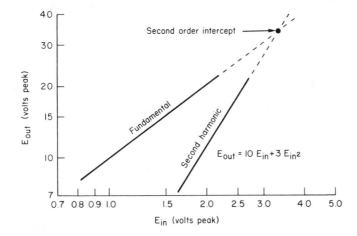

Figure 1.14 Relative size of fundamental and second harmonic outputs will depend on the size of the input signal.

the new signals are not musically related to the original tones. But, properly used in a communications circuit, this characteristic can be useful. Figure 1.15 shows the input and output spectrums of an amplifier with two input signals and a second-order characteristic.

This characteristic makes it possible to amplitude modulate signals, build mixer circuits for transmitters and receivers, and automatically change the gain of high-frequency amplifiers without causing serious problems. All of these will be covered in later chapters.

What happens if there are a large number of input frequencies present at the same time? The result isn't as messy as it might seem. To find the different outputs, just write down the original signals and then their second harmonics. Next, pick the inputs in pairs and write down the sum and differences for each pair. That's all there is. If third-order characteristics are involved, many more signals would have to be added and this is the reason that third-order distortion is so much more serious.

For example, if an amplifier with second-order distortion has four input signals at 1.0, 1.5, 9.8, and 10.0 kHz then the outputs will be at:

Inputs (kHz)	*Outputs* (kHz)	
1.0	1.0	
1.5	1.5	Originals
9.8	9.8	
10.0	10.0	
	2.0	
	3.0	Second Harmonics
	19.6	
	20.0	

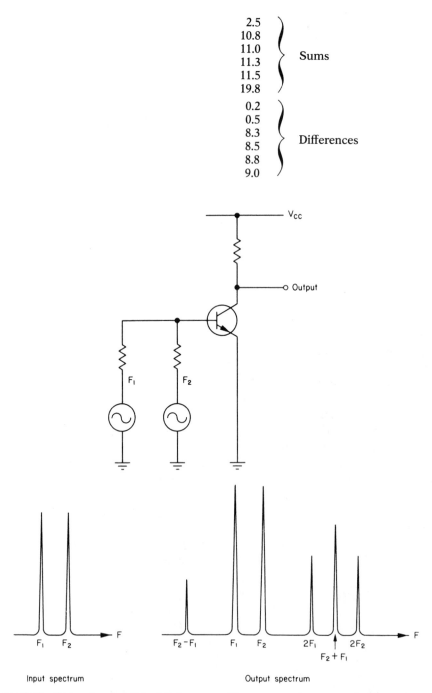

$$\left.\begin{array}{l} 2.5 \\ 10.8 \\ 11.0 \\ 11.3 \\ 11.5 \\ 19.8 \end{array}\right\} \text{Sums}$$

$$\left.\begin{array}{l} 0.2 \\ 0.5 \\ 8.3 \\ 8.5 \\ 8.8 \\ 9.0 \end{array}\right\} \text{Differences}$$

Figure 1.15 Second-order distortion in an amplifier can produce new signals at the sum and difference frequencies.

1.2.7 Phase and Delay Distortion

One form of distortion still exists. Even though a circuit is carefully designed to minimize random noise and to pass all important frequencies with equal amplitude and have absolutely linear amplitude characteristics, a waveform may still pass through badly distorted. The problem is delay distortion; if all frequency components of a waveform do not pass through a circuit with the same time delay, the output will not look the same as the input. This is illustrated in Figure 1.16 for the first three frequency components of a 100 Hz square wave.

   The initial waveform is shown in (a) along with its three frequency components drawn in proper time relationship. If this wave were to pass through a circuit that produced different time delays at different frequencies, the waveform at (b) might result.

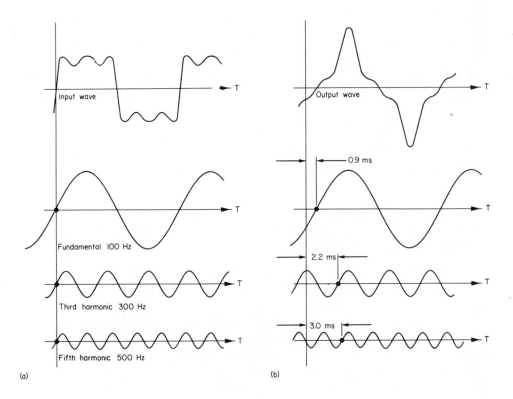

Figure 1.16 Distortion of a square wave that results from different time delays for each frequency component.

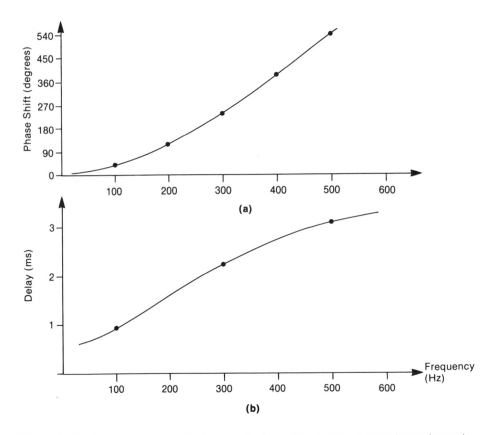

Figure 1.17 Phase response (a) of a circuit responsible for the distorted waveform of Figure 1.16(b). The calculated delay response is shown in (b).

The phase shift response of a circuit that could cause this distortion is shown in Figure 1.17 (a) and the corresponding delay is shown in (b). This delay is calculated for each frequency by,

$$\text{Time delay (seconds)} = \frac{\text{Phase shift (degrees)}}{360 \times \text{frequency (Hertz)}} \qquad (1.5)$$

The waveform of Figure 1.16 was unmodulated; no carrier was involved. When dealing with modulated waveforms, a slightly different delay is considered—the *envelope delay*. This describes the amount of time the information riding on the carrier takes to pass through a circuit. This can be measured by observing the input and output waves of the particular circuit on a good dual-beam oscilloscope as shown in Figure 1.18.

For delay measurements of any type, the absolute time that it takes for a signal to pass through a circuit is not usually as important as the relative varia-

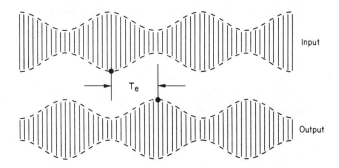

Figure 1.18 Measuring the envelope delay time required for a modulated sine wave to pass through a circuit.

tions in the delay for different frequencies. Referring back to Figure 1.17, this means that the values 0.9 ms, 2.2 ms, and 3.0 ms aren't as important as the fact that the delays at 300 and 500 Hz (2.2 ms and 3.0 ms) are respectively 1.3 and 2.1 ms greater than the delay at 100 Hz. This simply means that the horizontal axis of Figure 1.17 can be moved up to pass through the 100 Hz point. The vertical axis would now start at 0 ms at this point and would be relabeled *Relative Delay.*

The human ear is not particularly sensitive to delay variations—not until part of the signal starts to sound like an echo. A telephone voice channel, for example, has a rather wide delay variation across its 3 kHz band. For years, voice users experienced no problems but, with the increasing use of data transmission through the phone system, the variations must be smoothed out and extra charges are made for this "conditioning."

The picture on a TV set is an example of a signal that is very sensitive to delay distortion. Delay variations across the bandwidth of the video amplifiers will cause horizontal smearing on the screen. Also, the absolute delays of the black and white (luminance) amplifier and the color (chrominance) amplifier can cause the colors to appear at the wrong place on the screen if they are not identical—much like a bad printing job on the color comics in the newspaper.

All of these distortions will lead to some deterioration in the quality of the received signal. Frequency distortion, amplitude distortion, and delay distortion are most likely to occur in the transmitter and receiver circuits so a little extra care with their design will usually improve the quality of the communications. Internal noise is most serious in the first few stages of the receiver and, again, careful design will help. But not all of the distortions occur within the transmitters and receivers. The signal path can also be a great source of noise and delay distortion and these problems are much harder to control.

In spite of all the problems, our everyday life shows that quality AM, FM, and TV transmissions are possible thanks to careful planning and design.

1.3 RADIO FREQUENCIES

When carrier frequencies are modulated with information, a large increase in available transmission paths is obtained. But, what once seemed such a vast resource has rapidly become a very congested and limited range of frequencies that must be used with great care.

The available frequencies are divided into bands that tend to have some common characteristics, i.e., circuit techniques, noise levels, propagation characteristics. These bands are described in Table 1.1.

The wavelengths of the radio waves referred to in Table 1.1 are the distances that the signal travels while it oscillates through one complete cycle. This distance depends on the speed that the radio wave travels at and this, in turn, depends on the material that the wave is passing through—its dielectric constant in particular. Both air and a vacuum have approximately the same dielectric constant and so the speed of radio waves is 3×10^8 meters/sec., the same as light. (They are both electromagnetic waves.) The wavelength can be calculated:

$$\lambda \text{ (meters)} = 300,000/f(\text{kHz}) \tag{1.6}$$

$$\lambda \text{ (meters)} = \frac{300}{f(\text{MHz})} \tag{1.7}$$

$$\text{or, } \lambda \text{ (cm)} = \frac{30}{f(\text{GHz})} \tag{1.8}$$

where λ is wavelength
f is frequency

The VHF and UHF band references should be familiar to television set owners and the microwave X and K band designations may be familiar to automobile drivers in the habit of acquiring speeding tickets.[3]

When radiated from antennas, these radio waves could interfere with other users on the same frequency. Therefore, some agreement is needed, on a worldwide basis, about who uses which frequencies. This is decided every 20 years when the International Telecommunication Union (ITU) convenes a World Administrative Radio Conference (WARC).

The most recent meeting was in September of 1979 when some changes were made to the frequency allocations that will apply until the year 2000. For allocation purposes, the world is divided into three major regions as shown in Figure 1.19. Within each country, more detailed divisions of the frequency bands are made as required and licencing of individual stations is looked after.

---

[3] The microwave X and K bands are often used for police speed-detecting radar.

Table 1.1 Radio and microwave frequency bands (IEEE standard 521-1976).

Band	Frequency Range	Wavelengths
VLF	3 − 30 kHz	100 − 10 kilometer
LF	30 − 300 kHz	10 − 1 kilometer
MF	0.3 − 3.0 MHz	1000 − 100 meter
HF	3 − 30 MHz	100 − 10 meter
VHF	30 − 300 MHz	10 − 1 meter
UHF	300 − 1000 MHz	1 − 0.3 meter
L-band	1.0 − 2.0 GHz	30 − 15 cm
S-band	2.0 − 4.0 GHz	15 − 7.5 cm
C-band	4.0 − 8.0 GHz	7.5 − 3.75 cm
X-band	8.0 − 12.0 GHz	3.75 − 2.5 cm
K_u-band	12.0 − 18.0 GHz	2.5 − 1.67 cm
K-band	18.0 − 27.0 GHz	1.67 − 1.11 cm
K_a-band	27.0 − 40.0 GHz	1.11 − 0.75 cm
Millimeter	40.0 − 300 GHz	0.75 − 0.10 cm

Microwave {
L-band through K_a-band, Millimeter }

kHz	= kilohertz	1000 Hz
MHz	= megahertz	1,000,000 Hz
GHz	= gigahertz	1,000,000,000 Hz

In Canada, the licencing is the responsibility of the Federal Department of Communications (DOC); in the USA, it is the Federal Communications Commission (FCC), and in the UK, it is the British Post Office (BPO).

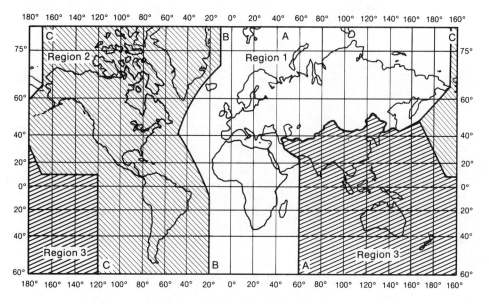

Figure 1.19 The three regions of the world for ITU frequency allocations.

For North America, which is part of ITU region 2, some of the common frequency users are shown in Figure 1.20. Notice that each line is a logarithmic scale; this tends to shrink the size of the higher frequency bands even though they contain a much wider frequency spectrum. The AM broadcast band, for example, is only 1060 kHz wide but appears, in the illustration, to be wider than the 420 MHz of the UHF TV band.

Figure 1.21 shows a photograph from the display of a spectrum analyzer. It is scanning the entire AM broadcast band (540 to 1600 kHz) and displays the

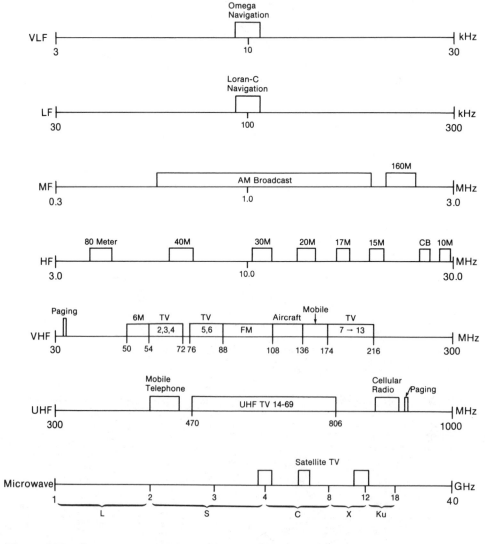

Figure 1.20   Common users of the radio frequencies up to 40 gigahertz.

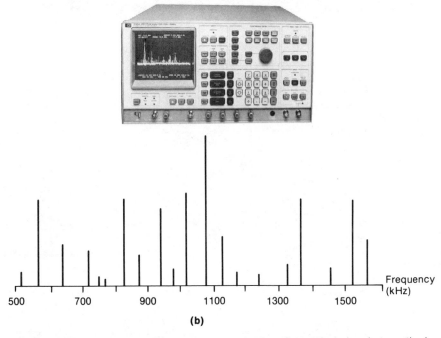

(b)

Figure 1.21 (a) A spectrum analyser and (b) its display of received signal strengths in the AM broadcast band.

strength of signals received near a large city. Each station occupies a 10 kHz range of the broadcast band as a result of the audio modulation of the carrier frequency. The signals for each station are centered around the assigned carrier frequency. Some stations are stronger than others. This could be due to their various distances away from the spectrum analyzer's location or to different transmitter power levels being used.

1.4 THE HISTORY OF RADIO

So far in this chapter, we have taken a look at some of the problems and solutions that must be considered by the designers of communication systems. This knowledge had to be gained one piece at a time over many years by a wide variety of people. Each built on the experience of others to reach today's level. The following paragraphs describe a few of the people and dates associated with these advances, but it should be remembered that hundreds of other people were involved in similar work and often earlier than those mentioned.

The history of communication must start with telegraphy over wires and then proceed to the development of wireless (radio) operations. In 1835, Sam-

International Morse Code

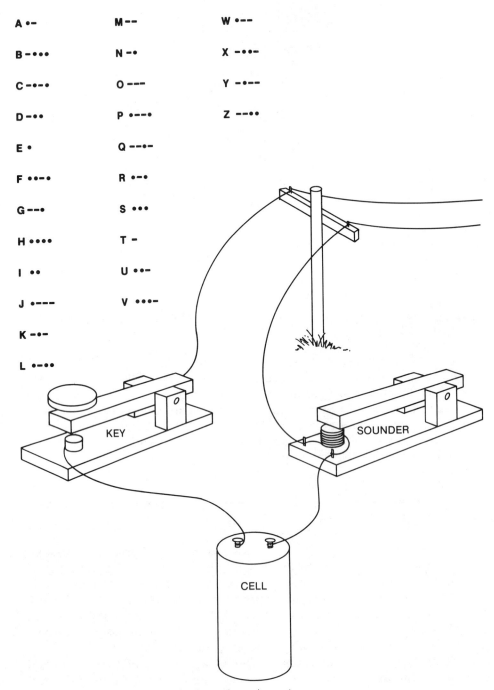

A •‒	M ‒‒	W •‒‒
B ‒•••	N ‒•	X ‒••‒
C ‒•‒•	O ‒‒‒	Y ‒•‒‒
D ‒••	P •‒‒•	Z ‒‒••
E •	Q ‒‒•‒	
F ••‒•	R •‒•	
G ‒‒•	S •••	
H ••••	T ‒	
I ••	U ••‒	
J •‒‒‒	V •••‒	
K ‒•‒		
L •‒••		

Figure 1.22 Morse's code and telegraph equipment.

uel Morse demonstrated a practical telegraph system capable of sending dots and dashes over a distance. The Morse code was developed to enable messages to be carried by the dots and dashes. By 1841, Morse had a commercial service operating between Washington and Baltimore. In 1858, a telegraph cable was laid across the Atlantic Ocean but was damaged several days later by using excessively high voltages. A permanent cable was completed several years later in 1866.

With telegraph lines successfully operating, the next step was to send voice directly over the wires and, in 1869, Alexander Graham Bell demonstrated his telephone. By now some information theory was becoming obvious; voice signals could not travel as far as the Morse code signals because they required a wider range of frequencies (about 4 kHz) and a higher signal-to-noise ratio to be properly understood. The transatlantic telephone cable had to wait for the development of a reliable electronic amplifier and it wasn't until 1956 that the telephone cable service was completed.

The existence of electromagnetic (radio) waves was shown theoretically by James Maxwell in 1864. By 1887, Heinrich Hertz had demonstrated in a practical way that it was possible to transmit and receive radio waves over a distance of 15 meters. His experiments used spark gaps resonant at about 80 MHz.

Then Marconi started his experiments. Much of the theory of radio already existed but Marconi acted as the "engineer" to put together a practical, commercially useful system of transmitters and receivers to carry telegraph signals. In 1897, he received a British patent for his system and, in the same year, set up the Wireless Telegraph and Signal Co. Ltd. By 1901, he had successfully sent Morse code across the Atlantic.

It is important to remember that all of the developments mentioned so far were accomplished without transistors or vacuum tubes. Circuits for Marconi's transmitter and receiver are shown in Figure 1.23. When the key was closed on the transmitter, a very high voltage pulse was produced at the secondary of the step-up transformer. As a result, an arc jumped across the gap between two metal balls.

An electrical arc contains electrons moving totally at random and so a noisy waveform similar to that of Figure 1.6 will be produced. This contains a wide range of frequencies and a selected group will be tuned, partly by the L and C indicated and partly by the length of the antenna wire. (It also has capacitive and inductive effects.) The antenna was a vertical wire and today vertical, grounded antennas are occasionally referred to as *Marconi antennas*.

The receiver resembles what would now be called a crystal set. No amplification was used, just a nonlinear element to "detect" or rectify the incoming signal. These early elements were called *coherers* and consisted, in some cases, of iron filings or a drop of mercury between two electrodes in a glass tube. The detected signal was then either listened to on earphones or fed to a sensitive "inker" that made dots and dashes on a moving paper tape. The sound heard in the earphones was a combination of hisses and crackles—the result of the elec-

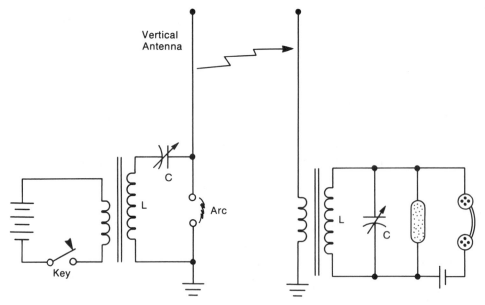

Figure 1.23 Marconi's (a) spark-gap transmitter and (b) receiver.

trical arc. And, so the signals occupied a wide bandwidth, making interference between stations a serious problem.

Finally, in 1907, Lee DeForest developed the triode vacuum tube and the new science of electronics started its growth to the huge technology it is today. With the triode, it was possible to build high-frequency sine wave oscillators for the transmitters with the result that the signals were confined to a much narrower bandwidth so that interference was drastically reduced. It also meant that amplifiers could be used for the weak signals at the receiver so ranges could be increased.

By World War II (1939–1946) most of the modern day analog circuits existed in vacuum tube form—the superheterodyne receiver, microwaves and radar, Hartley and Colpitts oscillators, cathode ray tubes, and both black and white and color television (although not in its final form).

More recent advances have seen the development of the transistor by Bell Laboratories in 1947, the planar-integrated circuit in 1959 at Fairchild, and the microprocessor in 1970 at Intel.

What about the future? With the advances in electronics in general and communications in particular that the author has seen, it looks foolish to even guess what will happen. But that is the purpose of this book; the students working through these pages are the ones who will be developing, servicing, and using the systems of the future.

1.5 REVIEW QUESTIONS

1. Name six different examples of communications systems.

2. Name the medium or path used in each of the above.

3. Must all communication systems use modulation?

4. What are some of the advantages gained by using modulation?

5. Is it possible to electronically separate the noise produced inside the receiver due to transistors and other components from external noise?

6. Which radio frequency band do each of these frequencies lie in? 7.9 GHz, 35 MHz, 809.4 MHz, 1600 kHz, 7.13 MHz, 212 kHz, 1.01 GHz.

7. Sketch the waveform transmitted by Marconi's spark-gap transmitter.

8. What are the four types of distortion that reduce the quality of communication signals?

9. Calculate the open-circuit white noise voltage coming from a video signal generator with a 60 Hz to 4.0 MHz bandwidth and with a source resistance of 75 ohms. It is a hot summer day and the temperature is 35°C.

10. What is the noise figure (in dB) of this satellite-TV pre-amplifier? The input signal from the antenna is 0.05 microvolts and the noise from the same source is 0.004 microvolts. At the output of the amplifier, the desired signal is 3.5 microvolts and the unwanted noise is 0.33 microvolts.

11. a. Draw a graph that represents the relationship between the input signal and the output signal for an amplifier with the transfer characteristic of:

$$E_{out} = 300 \ E_{in} - 25 \ E_{in}^2$$

(Use input limits of $E_{in} = -0.1$ to $+0.1$ volts.)

   b. Sketch the output waveform for a sine wave input that has peaks of $\pm$ 50 millivolts.

   c. Calculate the fundamental and second harmonic component of this output waveform.

12. List all the output frequencies that would result when these four signals are simultaneously fed to the input of an amplifier with a second-order transfer characteristic: 1.2, 4.5, 6.3, and 9.8 kHz.

13. Who are ITU, WARU, DOC, FCC, and BPO?

14. What distance will a radio wave travel in air in one microsecond?

15. What is the wavelength of radio signals at (a) 1430 kHz, (b) 19.2 MHz, and (c) 10.7 GHz?

16. What time delay (seconds) is created if a 3.19 MHz sine wave is shifted in phase by 203 degrees?

2. COMPONENTS

The major part of most communications systems is built up using individual resistors, capacitors, transistors, and integrated circuits that the student is already familiar with. But other, more mysterious elements are often included. Moreover, even the familiar capacitor and transistor may show unusual characteristics especially in the higher-frequency portions of transmitter and receiver circuits. The extreme of special techniques and components is reached at microwave frequencies where circuits resemble plumbing nightmares more than anything electronic.

This chapter will examine wire, inductors, capacitors, and transistors to see how they behave over a wide range of frequencies. The microwave work will be left to a later chapter.

2.1 WIRE

Wire is often used to connect different points in a circuit. It is also used to make components such as transformers and inductors. We will examine the characteristics of different types of wire first and then see what happens when the wire is coiled to form inductors.

Much of a wire's characteristics depend on the diameter used. Of course, all wire has some resistance of its own (not considering the low temperature superconductivity of some metals) and this resistance decreases as the diameter increases in size. Copper is the most common material used because it provides the lowest resistance of all metals with the exception of silver—and at a lower cost.

Wire diameters are commonly expressed as American Wire Gage (AWG) numbers. Table 2.1 shows the diameters for both bare and enamel-coated solid wires. The resistances shown are for soft-drawn copper wire and are only correct at the lower frequencies. Above a certain frequency, the resistance of all wire increases and this change, therefore, becomes the first of our odd effects.

Table 2.1 Wire diameter and resistance for wire gages 12-32 measured at 20°C. Litz wire is expressed as number of insulated strands of no. 44 wire.

Gage (AWG)	*Bare Diameter*		*Double Enamel-Coated Diameter*		*Resistance*	
	thousand	*mm*	*thousand*	*mm*	*per 1000 ft*	*per km*
12	80.81	2.052	83.8	2.13	1.67	5.488
14	64.08	1.628	67.4	1.71	2.614	8.576
16	50.82	1.291	53.8	1.37	4.646	15.24
18	40.30	1.024	43.1	1.10	6.693	21.96
20	31.96	0.812	34.6	0.879	10.46	34.30
22	25.35	0.644	27.6	0.701	18.58	60.97
24	20.10	0.511	22.2	0.564	26.77	87.82
26	15.94	0.405	17.8	0.452	40.81	133.9
28	12.64	0.321	14.4	0.366	64.89	212.9
30	10.02	0.255	11.6	0.295	103.2	338.5
32	7.95	0.202	9.5	0.241	164.1	538.3
5 × 44 Litz	nylon-wrapped		8.0	0.203	530.4	1740
6 × 44 Litz	nylon-wrapped		9.0	0.228	415.0	1361
7 × 44 Litz	nylon-wrapped		10.0	0.254	331.0	1086

2.1.1 High-Frequency Resistance of Wire

At DC, charge carriers are evenly distributed through the cross-section of a wire. As frequency increases, these carriers move away from the center and so less of the wire is available for conduction. The wire's resistance, therefore, increases with frequency. This is called the *skin effect*, which is a result of the strong magnetic fields near the center of the wire that increase the local inductive reactance. The charge carriers find it easier to move near the edge where the magnetic fields are a bit weaker.

When AC current is flowing, the charge carriers distribute themselves across the diameter of the wire as shown in Figure 2.1. The electrons have the greatest concentration at the wire's edge and least at the center. The *skin depth* is the distance in from the edge to a point where the number of electrons has dropped to $1/\varepsilon$ ($\varepsilon = 2.7183$) = 36.8% of the number at the edge. This depth conveniently lets us assume that all of the charge carriers flow uniformly in the outer "skin depth ring" and none flow inside. For copper,

$$\text{Skin depth } (\delta) = \frac{66.4}{\sqrt{F}} \text{ (mm)} \tag{2.1}$$

where F = frequency in Hz

The AC resistance of a wire at any particular frequency can be calculated by comparing the cross-section area of the "skin depth ring" to the full cross-

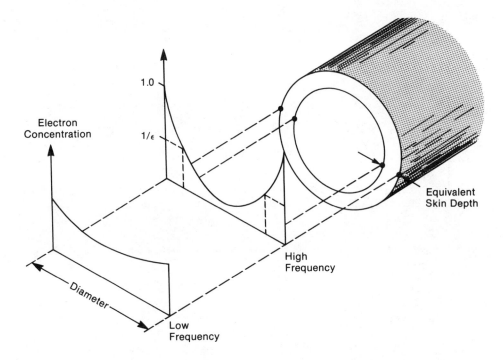

Figure 2.1 High-frequency skin effect causes charge carriers to move near the edge of a wire. The effective thickness of this layer is the skin depth.

section of the wire. As long as the skin depth is smaller than the wire's radius, the AC resistance will be:

$$R_{AC} = R_{DC} \times \frac{\text{total (DC) area}}{\text{ring (AC) area}} \qquad (2.2)$$

The general change of a wire's resistance with frequency due to "skin effect" is shown in Figure 2.2. At low frequencies, the curve is flat at the DC resistance value; at higher frequencies, it climbs in proportion to the square root of the frequency being used.

The curve bends at what could be called a *corner frequency* and, at this point, the AC resistance is 33.3% higher than the DC resistance. This corner frequency will change with different diameter wires and so is listed in Table 2.2 along with the DC resistance for each wire.

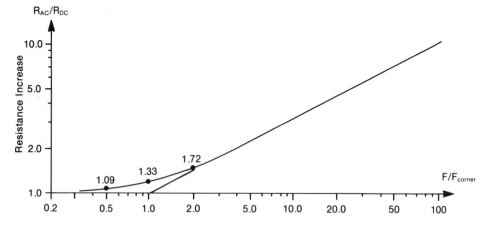

Figure 2.2 General change of a wire's resistance with frequency due to "skin effect."

Table 2.2 Corner frequencies for the high-frequency resistance of wire. To be used with Figure 2.2.

Gage (AWG)	DC Resistance (Ω)		Corner Frequency (kHz)
	per 1000 ft	per km	
12	1.67	5.488	16.61
14	2.614	8.576	26.38
16	4.646	15.24	41.95
18	6.693	21.96	66.65
20	10.46	34.30	106.0
22	18.58	60.97	168.5
24	26.77	87.82	268.0
26	40.81	133.9	426.0
28	64.89	212.9	678.0
30	103.2	338.5	1075.0
32	164.1	538.3	1710
34	260.9	855.96	2730
36	414.8	1360.9	4340

Example 2.1

By calculating the cross-sectional areas, find the AC resistance of # 22 copper wire at a frequency of 10 MHz.

Solution:

The DC resistance of this wire is 60.97 Ω per km and its diameter is 0.644 mm. The wire's total cross-section area is:

$$\text{DC area} = \pi r_1^2 = \pi \times \left(\frac{0.644}{2} \right)^2$$

$$= 0.3257 \text{ mm}^2$$

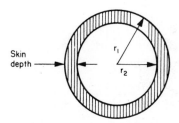

Skin depth

r_1

r_2

Figure 2.3

At 10 MHz, the skin depth is:

$$= \frac{66.4}{\sqrt{F}} = \frac{66.4}{\sqrt{10 \times 10^6}}$$

$$= .0210 \text{ mm}$$

The area of the skin depth ring (shaded area in Figure 2-3) is:

$$\text{AC area} = \text{total area} - \pi r_2^2$$

$$= 0.3257 - \pi \left(\frac{0.644}{2} - 0.0210 \right)^2$$

$$= 0.3257 - 0.2846$$

$$= 0.0411 \text{ mm}^2$$

The AC resistance at 10 MHz will be:

$$R_{AC} = R_{DC} \times \frac{DC\ area}{AC\ area}$$

$$= 60.97 \times \frac{0.3257}{0.0411}$$

$$= 60.97 \times 7.931$$

$$= 483.5\ \Omega\ per\ km$$

Example 2.2 ══

Using the standard AC wire resistance curve, find the resistance of 50 meters of #28 wire at 2.0 MHz.

Solution:

From Table 2.2, the DC resistance of 1000 meters of this wire is 212.9 Ω, so for 50 meters, the DC resistance would be 10.65 Ω. The corner frequency for this wire is 678 kHz. The 2.0 MHz frequency is

$$\frac{2.0\ MHz}{678\ kHz} = 2.950$$

times the corner frequency so this is the number we use on the horizontal scale of Figure 2.2. From the graph, we can read the resistance ratio,

$$\frac{R_{AC}}{R_{DC}} = 2.0$$

The resistance at 2.0 MHz is then

$$R_{AC} = R_{DC} \times 2.0$$

$$= 10.65 \times 2.0$$

$$= 21.3\ \Omega$$

2.1.2 Litz Wire

The information in Figure 2.2 and Table 2.2 can be used to draw some interesting graphs. Figure 2.4 shows the resistance changes of 1000 meter lengths of

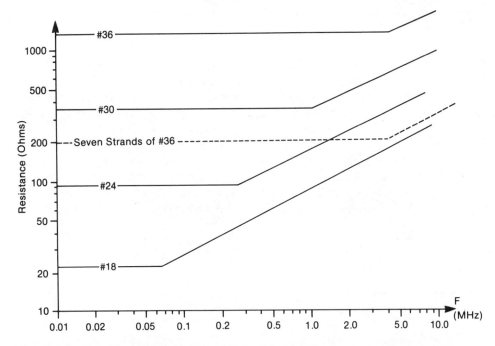

Figure 2.4 The AC resistance of nos. 18, 24, 30, and 36 wire and a bundle of seven lengths of #36 wire in parallel (Litz wire).

four different wire diameters. The interesting part is that, even though the thinner wire (#36) has a rather high DC resistance, it maintains this resistance up to a higher frequency than any of the other wires. What would happen if we put seven insulated strands of #36 in parallel? The dotted line on Figure 2.4 shows the result. We have a lower resistance and it stays constant over almost the same wide frequency range as the original wire. This frequency range is about seven times wider than that of a single wire with the same DC resistance.

A man named Litzendraht developed this wire and an abbreviated form (Litz) of his name is now used to identify the wire. Two precautions must be observed when it is manufactured:

1. The individual strands must be kept insulated from each other except at the two ends.

2. The strands must be woven so that each spends equal time in the center of the bundle, otherwise some strands will experience stronger magnetic fields and not carry their share of the current.

Litz wire is commonly used to make inductors and transformers that operate below 2 MHz. Three commercially available Litz wires are described in Table 2.1.

2.1.3 Inductance of Wire

Even a straight piece of wire will have some inductance since a magnetic field surrounds all current-carrying conductors. This is why we have been using the word *inductor* when referring to the component with all the wire coiled up inside. This coiling concentrates the magnetic field so that each turn is exposed to the total field. As a result, the inductance is increased beyond what it would have been if the same amount of wire were pulled out straight.

Straight wires are often used to interconnect various points in a circuit and so the self-inductance should be minimized. Since the inductance of a given length of wire increases as the wire gets thinner, the thickest wire possible should be used for interconnection where inductance could be a problem. The inductance of various lengths of four different wire sizes is shown in Figure 2.5.

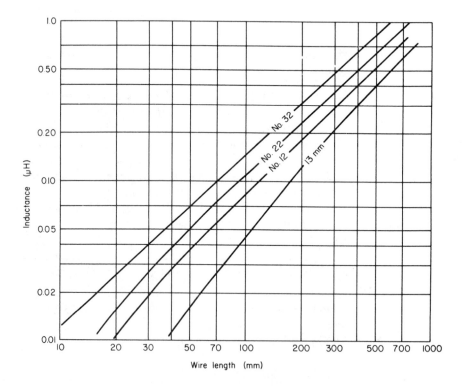

Figure 2.5 Low-frequency inductance of straight lengths of wire of various sizes.

Example 2.3 ═══════════════════════════════════════

In a CB set at 27 MHz, a 250 mm (about a 10-inch) length of #22 wire is used to connect a 50Ω source to a 50Ω load. Is this a simple connection or does the wire alter the circuit's characteristics?

Solution:

From Figure 2.5, the 250 mm length of #22 wire will have an inductance of about 0.30 μH. At 27 MHz, this inductance will have a reactance of:

$$X_L = 2 \pi FL$$

$$= 2 \pi \ 27 \times 10^6 \times 0.30 \times 10^{-6}$$

$$= + j \ 50.89 \ \Omega$$

This is moderately significant because it will increase the total impedance of the circuit from 100 Ω ($R_S + R_L$) up to:

$$Z_T = \sqrt{100^2 + 50.89^2}$$

$$= 112.2 \ \Omega$$

(a 12.2% increase)

This impedance increase will cause a corresponding drop in the current and power reaching the load (1 dB).

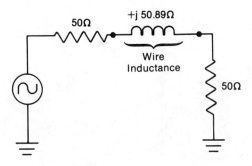

Figure 2.6

2.2 INDUCTORS

The purpose of this chapter is to look at practical components and, since wire is the main ingredient of inductors, we are well on our way to understanding practical inductors.

Since wire has a resistance that changes with frequency, we know that inductors will always have losses and these losses will increase at higher frequencies due to the skin effect. To describe the relative seriousness of the resistance, the term *component Q*, a short form of "component Quality," is used.

$$\text{Inductor } Q = \frac{X_L}{R_{series}}$$

Figure 2.7

The term is more specifically called *Inductor Q* to distinguish it from a similar description of capacitors to be discussed next.

Both the reactance of the inductance and its series loss resistance will change with frequency so the Q will also be frequency-dependent. A typical variation of Inductor Q with frequency is shown in Figure 2.8. The variation of the total impedance of the inductor with frequency is also shown.

The Inductor Q rises initially at the same rate as the frequency changes and this continues as long as the series resistance remains at the DC value. Then, at some frequency that depends on the wire diameter and also on the number of windings, the skin effect sets in and the series resistance starts to

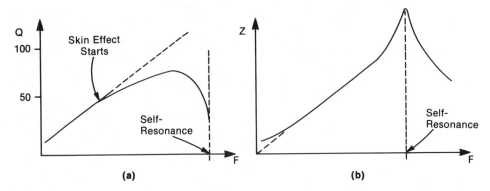

Figure 2.8 Typical variation of an inductor's Q (a) and impedance (b) with frequency.

climb. However it doesn't climb as fast as the frequency does (look again at Figure 2.2) and so the Q continues to rise but not as steeply. Then another problem sets in and the Q drops rapidly.

The problem is that all inductors are made with turns of wire that pass close to each other and this allows stray capacitance to build up. Along with the inductance, a parallel resonant circuit is formed and the resulting resonant frequency finishes the useful range of the inductor. This capacitance can be minimized with special construction methods but it can not be totally eliminated.

Example 2.4 ═══

Calculate the Q of a 1.5 mH inductor at both 100 and 250 kHz. Its series resistance is constant at 20 ohms for both frequencies.

Solution:

The reactance of the Inductor is:

At 100 kHz

$$X_L = 2\pi fL = 2\pi \times 100 \times 10^3 \times 1.5 \times 10^{-3}$$

$$= 942.5 \ \Omega$$

At 250 kHz

$$X_L = 2\pi fL = 2\pi \times 250 \times 10^3 \times 1.5 \times 10^{-3}$$
$$= 2356 \ \Omega$$

The Q is:

At 100 kHz

$$Q = \frac{X_L}{R_s} = \frac{942.5}{20} = 47.1$$

At 250 kHz

$$Q = \frac{X_L}{R_s} = \frac{2356}{20} = 118$$

Example 2.5 ━━━━━━━━━━━━━━━━━━━━━━━━━━━━━━━━━━━━━

Find the total impedance of this inductor at 5.0 MHz. It has a Q of 75 at this frequency and the stray capacitance of its windings is 6.2 pF

Solution:

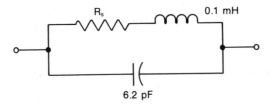

Figure 2.9

First we must find the reactance of the inductance at 5.0 MHz.

$$X_L = 2\pi fL = 2\pi \times 5.0 \times 10^6 \times 0.1 \times 10^{-3}$$
$$= + j\, 3142\ \Omega$$

Now we can use this to find the value of the series resistance (ignore the capacitance for the moment).

$$R_{series} = \frac{X_L}{Q} = \frac{3142}{75} = 41.89\ \Omega$$

Next the reactance of the winding capacitance is calculated at 5.0 MHz.

$$X_c = \frac{1}{2\pi fC} = \frac{1}{2\pi \times 5.0 \times 10^6 \times 6.2 \times 10^{-12}} = -j\, 5134\ \Omega$$

Finally, we put all of these together to find the total impedance. Remember that we are working with resistances and reactances that have three different vector angles.

$$Z_T = \cfrac{1}{\cfrac{1}{-jX_c} + \cfrac{1}{R_s + jX_L}}$$

$$= \cfrac{1}{\cfrac{1}{-j\, 5134} + \cfrac{1}{41.89 + j\, 3142}}$$

$$= \frac{1}{(+ j\, 1.948 \times 10^{-4}) + (4.243 \times 10^{-6} - j\, 3.182 \times 10^{-4})}$$

$$= 278.4 + j\, 8094 \; \Omega$$

or $8099 \; \underline{/88.0°}$

Comment: Notice that, in the initial statement of the problem for this example, the Q was given as 75. This statement ignores the presence of the winding capacitance and simply calculates the Q as equal to X_L/R_s. However, the solution to the example gives a final value, in rectangular form, consisting of a resistance of 278.4 ohms in series with an inductive reactance of $+j\, 8094$ ohms. The Q calculated from these values is:

$$8094 \,/\, 278.4 = 29.1$$

The huge drop from the first value of 75 down to 29.1 is caused by the presence of the parallel capacitance. (It was ignored when the value of 75 was calculated). The 5.0 MHz used for this example is very close to the resonant frequency of the inductor and its stray winding capacitance (6.4 MHz). This drop in Q was shown in Figure 2.8(a). The initial value of $Q = 75$ would be a point on the dashed line of this illustration. If the problem had been calculated at the exact resonant frequency then only a resistance would be found—no inductive reactance—and so the Q would have been zero.

For students that don't feel comfortable working with vectors, the same problem can be worked in a simpler manner as we will see in the next chapter.

Some typical coils are shown in Figure 2.10. Some of these are wound with many turns of closely-spaced Litz wire. This provides large-value inductances and high Q at the lower frequencies (below 2 MHz). For higher frequencies, smaller inductances are needed and so the coils have fewer turns of wire. But now, to keep the Q high, large wire diameter must be used or skin effect will become a serious problem. Also, the turns must be spaced wider apart to decrease the capacitance between each turn. This is necessary to keep the resonant frequency of the coil higher than the operating frequency.

2.3 INDUCTORS WITH MAGNETIC CORES

Small values of inductance can easily be obtained by winding a few turns of heavy wire around a pencil or drill bit and then slipping the coil off. If the wire is stiff enough, the resulting inductance can usually be soldered directly into a

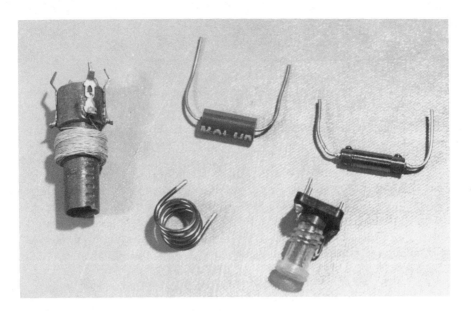

Figure 2.10 An assortment of low- and high-frequency inductors.

circuit without further support. For larger values, some support will be needed and so cardboard, plastic, or ceramic coil forms are used. These forms have no magnetic properties so the coil inductance depends only on the characteristics of the windings themselves.

However, for values larger than a few millihenries, the number of turns needed will often become excessive so magnetic cores are added. Materials such as laminated steel, powdered iron, and ferrites will increase the inductance of any coil of wire because they offer a lower resistance path to the magnetic lines of force and so concentrate the magnetic field closer to the wires. This is shown in Figure 2.11.

The use of magnetic cores can cause problems, however. One problem is that the material is not linear. What this means is that an inductance that uses a magnetic core material will have certain characteristics when a 1 volt signal is used and could have slightly different characteristics if a 2 volt signal is used. The problem is that all core materials have a "saturation point" beyond which they cannot handle any stronger magnetic fields. A good example of this problem is found in the "crossover networks" used in high fidelity speaker systems. Here large inductance values are needed and magnetic cores would be ideal to save wire. However, the nonlinearities of the core material would cause a very slight distortion in the sound as the signal level changed up and down. Of course, for a quality music system, this would not be tolerated and so only air is used for the core.

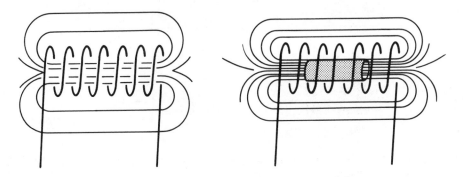

Figure 2.11 Magnetic cores concentrate the magnetic field around the turns of a coil more than an air core does.

The second problem with the core material is that it is frequency-sensitive. This means that if you can design a good coil at 1 MHz with a particular core, there is no guarantee that the results at 10 MHz will be any good at all. Usually a different type of material would be needed at the higher frequency.

The solution to both of these problems is either to work carefully with information sheets from manufacturers or to anticipate problems and carefully check any coils you make with magnetic core materials using both the correct frequency and signal level.

Magnetic cores also make it possible to increase the Q of an inductor. This may seem odd because the core material has losses of its own. The Q increase is possible because the use of the core means that less wire is needed and this decreases the total amount of resistance in the winding. As long as the losses in the core material are lower than what would have been lost in the amount of wire eliminated, the Q of the coil will increase. Of course it is a problem for the coil designer to see that this happens and not the reverse. An assortment of typical core materials and shapes is shown in Figure 2.12.

Figure 2.12 An assortment of magnetic cores used to increase the inductance of coils.

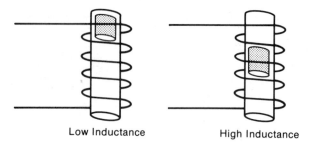

Low Inductance   High Inductance

Figure 2.13 The position of the tuning slug and its size determine how much of the coil's inductance is changed.

2.3.1 Variable Inductors

The use of magnetic cores makes it possible to conveniently change the inductance of a coil by moving a small "slug" in and out of the middle of the winding. The amount of the inductance change depends on the material used for the slug and the relative size of the winding and the slug. A small slug may only cause a 20% change in the coil's inductance while much larger adjustable cores that almost totally surround the winding may cause a 2:1 change.

A common use for variable inductances is found in a typical automotive radio. The AM broadcast band from 550 to 1600 kHz is tuned with the set of coils shown in Figure 2.14. The three are all mechanically connected so that the cores move in and out as the tuning knob is turned or one of the preset pushbuttons is pressed.

Figure 2.14 The tuning section of an AM automotive radio. The three cores of the coils are mechanically linked so that they move in and out together.

2.3.2 Inductor Measurement

Any measurement of an inductor should take into account the two separate parts of an inductor's equivalent circuit—its inductance and its resistance. And, since the resistance is likely to change with frequency due to the skin effect, the measurement should be made at the frequency of interest. A convenient meter that will do this is shown in Figure 2.15(a) along with a greatly simplified equivalent circuit. The internal oscillator can be set to any frequency within the range 20 kHz to 70 MHz. The variable capacitor is then adjusted to give a peak reading on the voltmeter. Calibration marks on the capacitor dial can be used to find the inductance and the peak reading on the meter indicates the Q and, therefore, the series resistance.

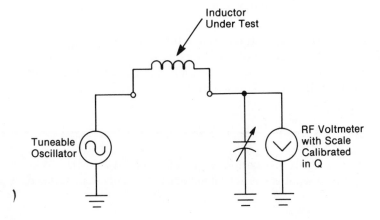

Figure 2.15 Hewlett-Packard 4342A Q meter and its simplified equivalent circuit. (Courtesy of Hewlett-Packard)

2.4 CAPACITORS

As with inductors, there are also losses in all practical capacitors and there is also a resonance problem. The energy losses are mainly due to the dielectric material used as insulation between the two plates of the capacitor. Since many different types of dielectric materials are used there are, therefore, many different capacitor qualities available. There are also corresponding variations in tolerances, temperature coefficients, and prices.

Capacitor losses are again treated as a series resistance and the term *component Q* still applies

$$\text{Capacitor } Q = \frac{X_c}{R_{series}}$$

Another term that is often used to describe capacitor losses is *dissipation factor*, where:

$$\text{Dissipation factor} = \frac{1}{\text{Capacitor } Q}$$

Whereas most inductors have Qs up to only 150, very few capacitors are this bad. Many, in fact, have Qs higher than 1000, indicating very small losses. Moreover, the Qs do not have the wide variation with frequency that inductors do.

Capacitors also suffer from a self-resonance problem in a manner somewhat similar to inductors. With capacitors it is the added series inductance of the leads and the internal electrodes. Keeping the leads short when soldering a capacitor into a high-frequency circuit can, therefore, make an important difference.

The variation of a capacitor's impedance with frequency is shown in Figure 2.17. The normal decrease of reactance with frequency suddenly steepens and, near resonance, the impedance is much lower than normal. The capacitor acts as if it had a larger value over a narrow frequency range.

Example 2.6 ━━━━━━━━━━━━━━━━━━━━━━━━━━━━━━━━━━━━━━━

A .001μF capacitor has a dissipation factor of .00125 at 1.5 MHz. What is the Q of the capacitor and what is the actual value of the equivalent series resistance?

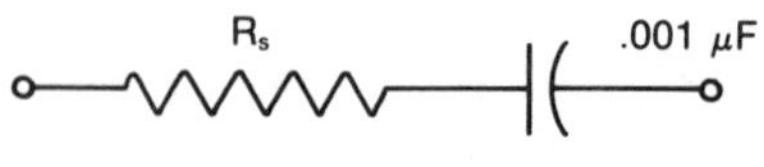

Figure 2.16

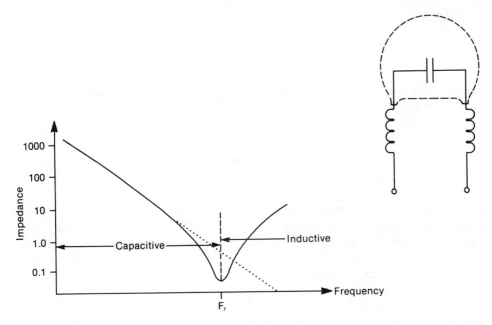

Figure 2.17 Variation of capacitor impedance with frequency. The series resonance is caused by the lead and electrode inductance.

Solution:

The Q of this capacitor is found by inverting the dissipation factor.

$$Q = \frac{1}{DF}$$

$$= \frac{1}{.00125}$$

$$= 800$$

To find the value of the series resistance we first need the reactance of the capacitor at 1.5 MHz.

$$X_c = \frac{1}{2 \pi F_c}$$

$$= \frac{1}{2 \pi \, 1.5 \times 10^6 \times .001 \times 10^{-6}}$$

$$= 106.1 \ \Omega$$

The series resistance is then:

$$R_s = \frac{X_c}{Q} = \frac{106.1}{800} = 0.133 \; \Omega$$

2.5 TRANSISTORS

A very important component of any electronic circuit is the transistor. It may appear as individual, discrete devices or as multiple transistors inside integrated circuit packages—for analog or digital functions. In any case, a knowledge of transistor operation and limitations is necessary in understanding communication circuits and systems. The student will likely have already completed at least one course in transistor circuits. However, a quick review will be given first, followed by an introduction to changes that occur at higher frequencies.

2.5.1 Low Frequency Operation

Transistors come in two basic types—bipolar and field effect. Both are used in communication circuit applications in either discrete or integrated circuit form. Each type is available in two polarities so that NPN and PNP bipolar transistors are used as well as N- and P-channel field effect types. While all four variations are used at the lower frequencies, there is a definite preference at higher frequencies for the NPN bipolar and the N-channel field effect transistors, the reason being the higher mobility of the charge carriers in these transistors. This results in higher gains at the higher frequencies.

Transistors of any type can be operated as either linear or nonlinear amplifiers, depending on the biasing used and the size of the input signals. It may seem odd to intentionally design circuits that would intentionally cause distortion but there are a number of circuits where this is necessary. Mixer circuits and variable gain (AGC) amplifier circuits are two examples from radio receivers, and frequency multipliers and power amplifiers are examples from transmitters. Logic gates are an example of extreme nonlinearity since they usually operate in only two states—full on and full off.

For the introductory portion of our transistor discussion, we will concentrate on the NPN bipolar transistor. The bipolar transistor will only operate if its collector-base junction is reverse biased by having the collector for the NPN transistor biased several volts more positive than the base voltage. With only

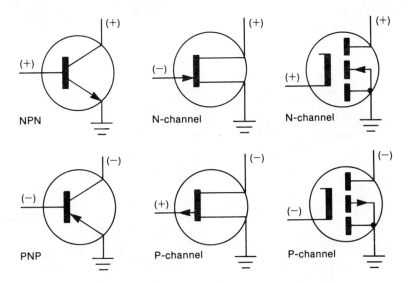

Figure 2.18 Symbols used for transistors. (a) bipolar transistors, (b) junction FETs, and (c) metal-oxide-semiconductor transistors (MOSFET's). The polarities relative to ground show the normal voltages needed for linear operation.

this voltage present, no collector current will flow because of the reversed junction. The 12 volt power supply shown with the transistor in Figure 2.19 supplies this reverse bias and, with no additional connections, the transistor will be cut off which means that no current will flow. The collector-emitter voltage will be almost the same as the power supply voltage. For a logic circuit, this would be called a Hi output.

To get the transistor operating, a forward-bias voltage must be added to the base-emitter junction. For a silicon transistor, this is usually about +0.6 volts. The forward bias causes electrons to start flowing from the emitter region into the base region. However, they don't all go out the base wire. If the collector voltage is high enough (anything more than 1.0 volts) then most of the electrons will head straight up to the collector region and only about 1% will go out the base wire. There are two points that allow this to happen. First, the base region must be very thin so that the electrons have little chance of being "caught." Second, the NPN layers must be formed in a single crystal structure. This explains why you can't simply solder together two diodes as suggested by Figure 2.19 and have a working transistor.

For most modern transistors, about 99% of the emitter electrons find their way into the collector region and make up the external collector current. The remaining 1% or less come out the base lead and form the base current. If the base-emitter voltage is increased, more electrons will be attracted to the base lead and so the base current will rise (this is the relation shown in Figure 2.20). But, as the base current rises, so will the collector current. The base current

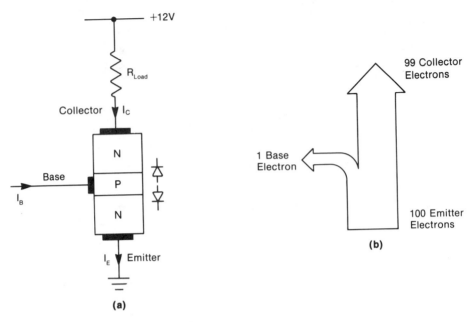

Figure 2.19 An NPN bipolar transistor is connected to a +12 volt power supply to reverse bias its collector-base junction (a). The diagram in (b) shows the relative size and directions of electron flow that will result once the transistor starts to operate.

seems to be controlling the collector current. Since the collector current is larger, the transistor seems to have a certain current "gain." The current gain is often called the *Beta* (β) of the transistor.

$$\text{Current gain } (\beta) = \frac{I_c}{I_B}$$

You will likely realize that the large number of extra electrons that are appearing at the collector are not being magically created by this gain. Rather, the small base current is simply controlling the flow of energy from the power supply. If you want 1 watt of signal out of the transistor then the power supply had better be capable of supplying at least this amount or the circuit won't work. The maximum value of collector current that can flow is set by the size of the supply voltage and the value of the series resistor. As the collector current rises, the increased voltage drop across the collector resistor will result in a reduction in the collector voltage. When the maximum value of collector current is reached, the collector-to-emitter voltage will be zero. The transistor will now be "saturated." Any further increase in base current under this condition results in no change in the collector current.

To design a transistor amplifier for high-power outputs (either audio or RF), high supply voltage is needed with a high current capability (the combination can be dangerous to careless finger poking!) and low values of load are also required.

Figure 2.20 is one of a series of graphs that can be used to describe the operation of transistors at lower frequencies. It shows that a forward-bias voltage of about 0.50 volts is needed before any base current can flow and that a small increase in voltage above this value will cause a rapid increase in base current. The curve also shows that the relationship is never linear—as the base current increases, the curve gets steeper.

This curved characteristic is one of the features used in the design of mixer circuits and variable gain amplifiers. These will be looked at in later chapters. For audio applications such as Hi-Fi and stereo, the curve is a nuisance since it causes unwanted distortion. Designers take great care and use such techniques as negative feedback to eliminate most of the problem.

Figure 2.21 shows a simple transistor amplifier and some input and output waveforms. The base is connected to a set of bias resistors and a coupling capacitor. This allows the base to sit at an average of about 0.60 volts and keeps some base current flowing all the time. As long as the input signal isn't too large, the output signal of the transistor will be about the same shape as the input, just larger. If the input signal is increased in amplitude, base current may not flow all the time and serious distortion of the output signal would result. Therefore, distortion or nonlinear operation depends on both the transistor's characteristics and the size of the input signal used.

Another piece of information that can be obtained from the input characteristic (Figure 2.20) is the input resistance of the transistor. This will tell us that, for a particular bias point, a small change in the base voltage will cause a certain change in the base current.

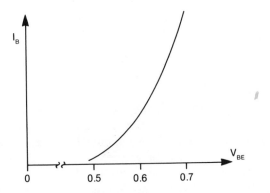

Figure 2.20 Input characteristic of a silicon bipolar transistor showing the relation between base-to-emitter voltage and base current.

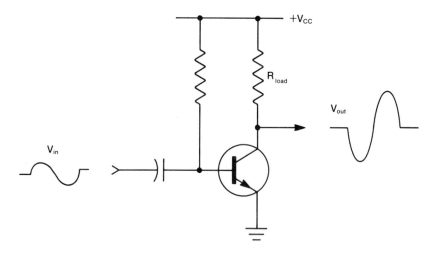

Figure 2.21 Biasing is added to the input of a transistor amplifier so that both the positive and negative halves of the input signal can be amplified.

$$\text{Input resistance} = \frac{\Delta\,V_{BE}}{\Delta\,I_B}$$

Because of the curved nature of the input characteristic, Figure 2.22 is the result. This shows that the input resistance of the transistor will decrease as the device is biased at higher values of current. Higher frequency operation usually requires these higher currents and so many communication circuits usually involve lower input resistances than would be found, for example, in audio circuits.

Again, this change in resistance can be useful in some circuits. You may have learned, in a previous transistor course, that the voltage gain of a single-stage amplifier similar to that of Figure 2.21 is given by:

$$A_v \approx \beta \times \frac{R_{\text{LOAD}}}{R_{\text{INPUT}}}$$

When the bias current through the transistor increases, the input resistance gets smaller and, as we will see shortly, the current gain (β) gets slightly larger. Both of these effects result in an increase in voltage gain. Therefore, a method of electronically controlling the gain of an amplifier is to simply change its bias point—more current, more voltage gain.

The curves of Figure 2.23 represent another important characteristic of transistors. They show the relation between I_b, I_c and V_{ce}. For collector voltages between roughly 1 and 30 volts, the collector current changes very little as long

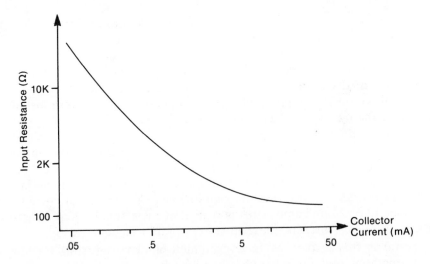

Figure 2.22   Input resistance of a typical bipolar transistor decreases as the average bias current increases.

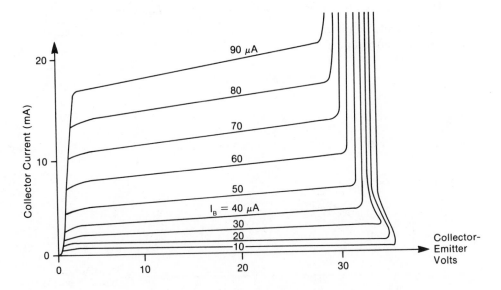

Figure 2.23   Collector characteristics of a bipolar transistor. Each curve represents the relation between collector current and collector-emitter voltage for constant values of base current.

as the base current is held constant. Above 30 volts (an arbitrary value that depends on the individual transistor) this particular transistor shows a rapid rise in current due to secondary breakdown—a condition that can destroy the transistor. This is mentioned here because some circuits using inductors, transformers, or tuned circuits can produce peak collector voltages several times higher than the power supply used. Careless operation could cause instant destruction of the transistor. Even more serious, careless finger poking could result in painful shocks (or worse).

The graph also shows that the vertical spacing between the curves increases as the collector current increases. This is the cause of the changes in current gain that have already been mentioned. This change is shown in more detail in Figure 2.24. Notice that the gain increase only occurs at currents less than about 10 or 20 milliamps (this is typical of most small signal transistors) and that, at higher currents, the gain starts to fall off again.

So far in this section, we have examined bipolar transistors for low-frequency operation and a number of points have been discussed that will be important to the understanding of communication circuits:

- Transistors are inherently nonlinear. This is good for some applications and bad for others, such as Hi-Fi.

- The input resistance gets lower as the transistor is biased at higher values of collector current.

- The current gain rises as the collector current increases.

- The collector-emitter voltage has a maximum value which, if exceeded even for an instant, could destroy the transistor.

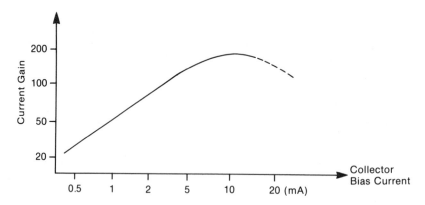

Figure 2.24 Current gain of a typical small-signal bipolar transistor changes with the value of collector current.

Much of this may have been review for you but, hopefully, you will remember some of the points mentioned. Now we will continue and describe what changes occur as a transistor operates at higher and higher frequencies.

2.5.2 Transistors at High Frequency

Figure 2.25 shows a normal bipolar transistor with three components added. This is approximately the condition that exists inside most general purpose transistors, although the values will change from one transistor type to the next. The capacitances are the result of the two semiconductor junctions: the collector-base and the emitter-base junctions. The collector capacitance is much smaller because the doping of this junction is lower and because the reverse-bias voltage at the collector pulls the charge layers further apart. The emitter capacitance is larger because of the heavier doping and the forward biasing that moves the base and emitter charge layers closer together. The base resistor appears because there is always a certain amount of silicon inside the transistor between the point of attachment of the base lead and the place where the transistor action occurs. Remember that these are not components that the transistor designer has added. Rather, the designer tries very hard to keep the values small because the smaller they are, the better the transistor will operate at higher frequencies.

What effect will this have on the operation of the transistor? At low frequencies such as audio there is very little effect. The capacitors will have a much higher reactance than the transistor so the added components can be ignored. As the operating frequency increases, the reactances drop and the problems begin. For most general purpose transistors, the effects start to show up at about 1 MHz. These effects consist of:

- Steadily decreasing current gain.

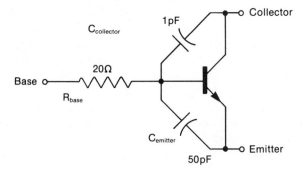

Figure 2.25 Three components can be added to a conventional transistor to explain what happens at higher frequencies.

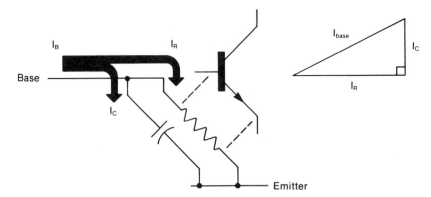

Figure 2.26 Input current to the transistor will follow two paths. Only the portion of the current that enters the active portion of the transistor (I_R) will be amplified.

- A lowering of the input resistance along with the appearance of a very reactive component.
- An increased tendency for signals to sneak back to the input from the collector of the transistor.

We will examine each of these problems briefly. First let's find out what happens to the current gain and why. Figure 2.26 shows that base current entering the transistor can follow two parallel paths. One is through the normal base-emitter junction of the transistor. This part of the input current will be amplified by the transistor. The other path is through the base-emitter capacitance. This shunts some of the input current away from the transistor and this portion will not be amplified. The result, as operating frequency increases, is that more of the input current is kept away from the active portion of the transistor and is not amplified. Therefore, the current gain of the overall device suffers.

The general shape of the current gain curve with changing frequency is shown in Figure 2.27. For this example, the gain starts to fall off at about 1 MHz which is quite typical of general purpose transistors. For the more general case, the gain will have fallen by 3 dB at the frequency where the capacitive reactance is the same value as the AC input resistance of the transistor itself. This frequency is called the *Beta cutoff frequency*; its formula is given in Example 2.7.

The Beta cutoff frequency isn't the upper frequency limit of the transistor's operation; it is just the frequency where changes start to occur. Above this frequency, the gain will of course be lower but it is only dropping at 6 dB per octave which isn't very fast. Therefore, the transistor continues to be useful well above the Beta cutoff frequency. The following example shows some of the simple calculations that can be made concerning current gain at higher frequencies.

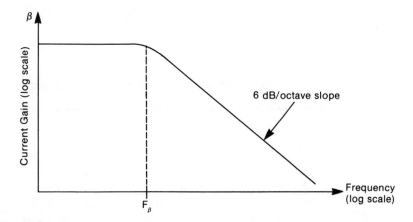

Figure 2.27 Variation of AC current gain of a general purpose transistor.

Example 2.7 ══════════════════════════════════════

A general purpose transistor has an emitter capacitance of 40 pF and an input resistance of 800Ω. Its current gain at very low frequencies is 170. What is its β cutoff frequency and what will the current gain be at 27.0 MHz?

Solution:

The β cutoff frequency is:

$$F_\beta = \frac{1}{2 \pi R_{in} C_e}$$

$$= \frac{1}{2 \pi \, 800 \times 40 \times 10^{-12}}$$

$$= 4.974 \text{ MHz}$$

Now to find the current gain at 27.0 MHz. If you look back at Figure 2.27 to see how the current gain changes with frequency, you will recall that the current gain drops at 6 dB per octave above the β cutoff frequency. This particular slope means that, as the frequency doubles, the current gain drops to $\frac{1}{2}$ and if the frequency triples then the gain will drop to $\frac{1}{3}$, etc. In other words, there is a linear inverse relationship. The current gain at 27.0 MHz is, therefore:

$$\text{Current gain} = 170 \times \frac{4.974}{27.0}$$

$$= 31.32$$

The transistor still has plenty of useful gain left at this frequency.

The small collector capacitance (usually less than 1 pF) also plays a part in the type of calculation we have just made. However, to determine how significant the effect would be, we would have to know the voltage gain of the circuit the transistor is used in and this isn't always an easy task. Figure 2.28 shows how the small collector capacitance acts as a much larger "Miller equivalent capacitance." If the voltage gain (A_v) of the circuit is known, then:

$$C_{equiv.} = C_{collector} (1 + A_v)$$

The reason for this increase in effective value is that any capacitance across the base-emitter junction has only the small input voltage across it. On

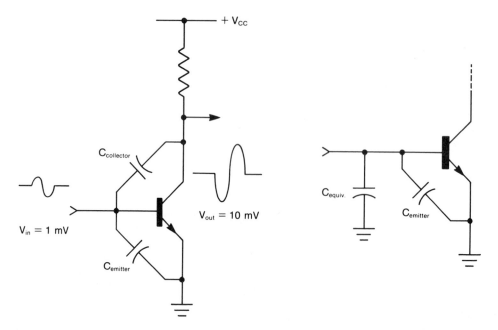

Figure 2.28 The small collector capacitance has 11 mV across it. The same portion of the input current will flow through it if a larger equivalent capacitance is placed base-to-emitter where the voltage difference is only 1 mV.

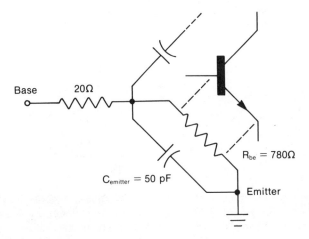

Figure 2.29 Components of the transistor responsible for variations of input impedance with frequency.

the other hand, any capacitance across the collector-base junction has the input plus the output voltage across it. (Remember that the input and the output will be out of phase.) This large voltage drop can result in a fairly large current even though the capacitance is small. The important point to remember is that a small collector capacitance can cause a fairly large loss of gain.

The next high-frequency effect we will look at is the input impedance change with frequency. At some frequencies, the impedance will be simply a resistive value while at other frequencies it will be a combination of resistance and capacitance. Figure 2.29 shows the parts responsible for these changes.

At very low frequencies, the input impedance of the above transistor will simply be the two resistors in series:

$$R_{in} = 20 + 780 = 800 \ \Omega$$

The capacitors will have too high a reactance to be noticeable. As the operating frequency increases, the emitter capacitance will slowly short out the 780 ohm base-emitter resistance. For this range of frequencies, the input impedance will be a combination of resistance and capacitance. At very high frequencies, the emitter capacitance will have an extremely low reactance and so the only part left is the 20 ohm series resistor. The input impedance will be just a resistive value once again.

This variation is shown in Figure 2.30. The first "corner frequency" at about 4 MHz is the frequency where the 50 picofarad capacitor and the 780 ohm resistor have the same value of reactance/resistance. The next corner on the curve is at about 159 mHz and occurs when the same capacitor and the 20 ohm resistor have the same value of reactance/resistance.

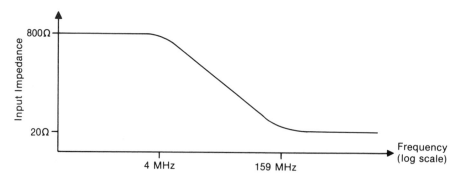

Figure 2.30 Variation of the input impedance of the transistor model shown in Figure 2.29. Both the low- and high-frequency impedances are resistive while the part in between is complex—both resistive and capacitive.

As with the current gain, the collector capacitance also plays a role in the input impedance. This can be found by calculating the equivalent Miller capacitance and then adding this value to the emitter capacitance. Again, the voltage gain must be known to do this. A point to remember is that the equivalent capacitance depends on what is connected at the collector of the transistor and so the input impedance can then change if changes are occurring at the collector such as might occur when tuning adjustments are made.

Field effect transistors also suffer from the same variation in impedance and the effect is more severe than for a bipolar. For audio applications, for example, a FET may have an input resistance of several hundred million ohms. But, at 100 MHz, the same transistor may have dropped to less than one thousand ohms. This is still a higher value than for the bipolar but the magnitude of the change is much greater.

Finally, we look at the instability of the transistor. This means that an amplifier might tend to oscillate because some of the output signal will sneak back from the collector to the base through the collector junction capacitance (or, in a FET, through the gate-drain capacitance). Whether oscillations will actually occur or not is difficult to predict but the tendency is always there. There are some fairly complex formulas that will indicate whether a design is stable or not but, for this text, the student is asked to consider the prime law of all technical designs. This law is of far greater importance than even Ohm's Law. It states that "anything that can go wrong will go wrong at the time least expected and at the point that will cause the most damage"—Murphy.

Why don't these oscillations show up in audio amplifier circuits? There are two reasons. First, the reactance of the collector capacitance will be so high at these low frequencies that very little signal will get back to the input. The second reason is that the maximum phase shift that can occur as a signal passes through a capacitor is only 90°. This isn't enough to make up for the 180° inversion that is produced by the normal forward gain of the transistor. In higher-frequency designs, tuned circuits are often used and these can add the additional shift necessary (unfortunately) to get oscillation going.

2.6 REVIEW QUESTIONS

1. Why is metal tubing often used for VHF and UHF circuits and antennas?

2. What is the advantage of Litz wire over normal stranded wire?

3. What are three factors that limit the practical frequency range over which an inductor may be used?

4. Why must a technician be concerned with component lead length when constructing a VHF circuit?

5. What circuit technique is used to reduce the distortion caused by the curved input characteristic of Figure 2.20?

6. Can a transistor be used at frequencies higher than its Beta cutoff frequency?

7. What effect does higher voltage gain of a transistor have on its input impedance?

8. What is the DC resistance of 650 meters of #28 copper wire?

9. What is the skin depth for copper wire at 1.5 MHz?

10. Calculate the AC resistance of 650 meters of #28 copper wire at 1.5 MHz.

11. If 50.0 cm of #32 wire is used to connect a 75 ohm source to a 125 ohm load, how much power will be lost at 15 MHz?

12. A 1.2 mH inductor has a constant series resistance of 15 ohms. Find the frequency at which this inductor will resonate with a 150 pF capacitor. What is the impedance at resonance and what is the bandwidth?

13. A 0.01 μF capacitor has a lead inductance of 0.08 μH. If the capacitor is used at 4.8 MHz, what will its capacitance "appear" to be at this particular frequency?

14. a. What is the Beta cutoff frequency of a transistor with an emitter capacitance of 15 pF and an emitter resistance of 1200 ohms?

    b. If the low-frequency current gain is 200, at what upper frequency does the current gain drop to 1.0?

15. Determine the input impedance (rectangular coordinates) of a transistor at 15.0 MHz for the following conditions: voltage gain = 20, base resistance = 12 ohms, emitter capacitance = 75 pF, collector capacitance = 1.3 pF, emitter resistance = 1000 ohms.

3. FILTER CIRCUITS

In this chapter we will look at the use of inductors, capacitors, and resistors to make filter circuits. Filters are used to get rid of unwanted frequencies while still retaining other, useful frequencies. Filters may also be used as impedance-matching networks to provide a more efficient transfer of power from one point in a circuit to another. The theoretical design of filters and how they work will be discussed along with a realistic look at practical designs that result when real-life components are used.

3.1 FILTER APPLICATIONS

First, we will look at two applications that require filtering to see what types and characteristics are needed.

For computer data transmission over telephone lines, it is convenient to produce a sine wave digitally. The resulting waveform is shown at the input to a filter in Figure 3.1(a). The stepped nature of the sine wave means that many additional frequencies are present in addition to the one that we want—the waveform's fundamental—and these show up in the corresponding frequency spectrum. After passing through the filter, the waveform looks much more like a sine wave and the frequency spectrum shows the unwanted harmonics to be greatly attenuated (reduced in amplitude).

Because only the lowest frequency passes through the filter and all higher frequencies are attenuated, this filter is called a *low-pass filter;* its general frequency response is shown in Figure 3.2. The response curve shows that a wide range of frequencies are passed with a little attenuation that is unfortunately always present—a characteristic of any practical design. This is the "passband" that extends up to the "cutoff frequency" where the attenuation suddenly starts to increase. Ideally, the attenuation should drop straight down here but, even if we had perfect components, this would be impossible. A slow increase in attenuation in the "stopband" is the result and the slope is often referred to as the

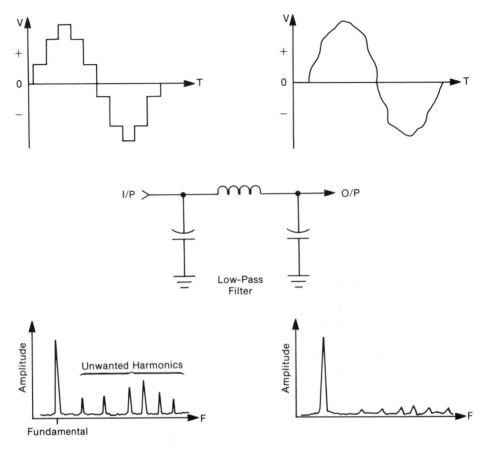

Figure 3.1 A low-pass filter is used to remove unwanted harmonics of a digitally synthesized sine wave.

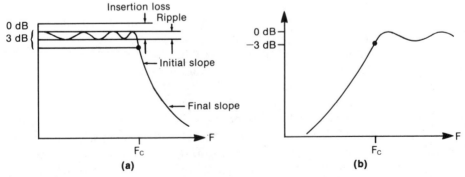

Figure 3.2 (a) the general amplitude response of a low-pass filter and (b) corresponding high-pass filter.

skirts of the filter response. Steep skirts are harder to design than shallower skirts.

A second situation that requires the use of a filter was shown in the spectrum analyzer display of Figure 1.20. To design a receiver for the broadcast band, we need a filter to select one of these signals and get rid of the unwanted signals at both higher and lower frequencies. In other words, what is needed is a *bandpass filter*. This must be able to pass the total signal from the one station we want to listen to; for AM stations this means a bandwidth of about 10 kHz. The general characteristics of bandpass filters are shown in Figure 3.3. Again we have the same problem with the skirts—shallower skirts are easier and cheaper to design for but steep ones are often needed. For AM station reception, the closest unwanted signals could be only 10 kHz above and below the station we do want and very steep slopes are required.

One often used description of a bandpass filter is *shape factor*, which is the ratio of the filter bandwidth at two different attenuation levels, usually 3 dB and 60 dB.

$$\text{Shape factor} = \frac{\text{bandwidth (60 dB)}}{\text{bandwidth (3 dB)}} \tag{3.1}$$

A shape factor of 1.0 would indicate vertical skirts and would be the ideal but unattainable bandpass filter. The best that can be done without getting too costly is a shape factor of about 1.2.

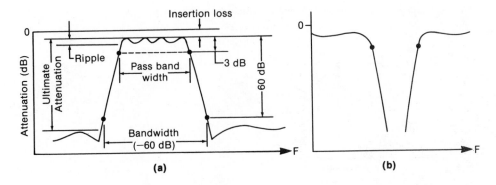

Figure 3.3 (a) the general amplitude response of a bandpass filter and (b) corresponding band-rejection filter.

3.2 SIMPLE LOW-PASS FILTERS

The simplest of all filters consists of one reactive element (one capacitor or inductor) as shown in Figure 3.4. In the positions shown, the circuits form low-pass filters[1] with identical gain and phase characteristics as shown in Figure 3.5.

The vertical attenuation scale of Figure 3.5 is calibrated in decibels and, therefore, is a logarithmic scale. The horizontal frequency scale is also logarithmic. With these scales we see that the attenuation curve gradually picks up a constant slope that is a straight line on the graph. The final slope is 6 dB for each octave change in frequency and this is very characteristic of any filter that uses one reactive element. The final phase angle change of 90° is also characteristic of the one-element design. A greater number of reactive elements will increase both the slope and the total phase shift.

Both the resistance of the source and load are shown in Figure 3.4 but notice that they are not considered to be part of the filter itself. However, the resistances play a very important part in the operation of the filters since any value changes would change the responses.

For the shunt capacitor circuit of Figure 3.4(a), the load voltage and its phase angle can be calculated from,

$$V_{\text{load}} = V_{\text{source}} \times \frac{R_L}{R_L + R_s} \times \frac{-jX_c}{R_t - jX_c} \tag{3.2}$$

where the total resistance,

$$\begin{aligned} R_t &= R_s \parallel R_L \\ &= \frac{R_s \times R_L}{R_s + R_L} \end{aligned} \tag{3.3}$$

[1]Low-pass filters using one reactive element are often known as *one-pole filters*.

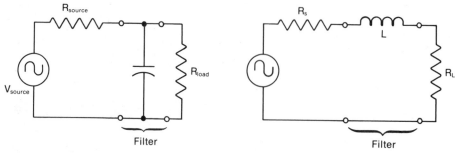

Figure 3.4 One-element, low-pass filters consisting of (a) a shunt capacitor or (b) a series inductor.

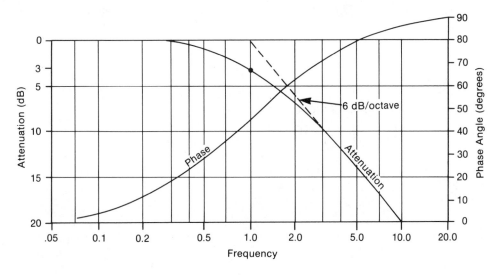

Figure 3.5 Attenuation and phase shift of single-element, low-pass filters.

Equation 3.2 contains two parts. The first part,

$$\frac{R_L}{R_L + R_s}$$

is a voltage divider ratio that does not change with frequency. It just says that the load voltage will always be less than the source voltage (unless the load resistor happens to be infinite).[2] The second part of the equation,

$$\frac{-jX_c}{R_t - jX_c}$$

changes in both magnitude and phase angle with changes in frequency. This is the part used to draw the two curves in Figure 3.5.

$$\text{Attenuation (dB)} = 20\log_{10}\left(\frac{X_c}{\sqrt{R_t^2 + X_c^2}}\right) \tag{3.4}$$

$$\text{Phase angle (degrees)} = \tan^{-1}\left(\frac{R_t}{X_c}\right) \tag{3.5}$$

[2] Having an infinite value load resistor on a filter may sound like a good idea since the load voltage will be maximum. But an infinite resistor will consume no power. Remember, $P = V^2/R$. For most applications, it is far more important to transfer as much "power" from the source to the load as possible and so R_s should equal R_L. When this is not possible, some filters can act as impedance-matching networks to improve the transfer.

The cutoff frequency of this filter is the point where the attenuation has reached 3 dB and is given by,

$$F_c = \frac{1}{2\pi R_t C} \text{ (Hz)} \tag{3.6}$$

Similarly, for the series inductor circuit of (b),

$$V_{\text{load}} = V_{\text{source}} \times \frac{R_L}{R_L + R_s} \times \frac{R_t}{R_t + jX_L} \tag{3.7}$$

$$\text{Attenuation (dB)} = 20 \log_{10}$$
$$\frac{R_t}{\sqrt{(R_t)^2 + (X_L)^2}} \tag{3.8}$$

$$\text{Phase angle (degrees)} = \tan^{-1}\left(\frac{X_L}{R_t}\right) \tag{3.9}$$

The cutoff frequency is,

$$F_c = \frac{1}{2\pi R_t L} \text{ (Hz)} \tag{3.10}$$

Now for an example to check our understanding.

Example 3.1 ▬▬▬▬▬▬▬▬▬▬▬▬▬▬▬▬▬▬▬▬▬▬▬▬▬▬▬▬▬▬

Using the circuit of Figure 3.4(a), calculate the cutoff frequency of the filter and its attenuation and phase shift at 250 KHz. The component values are:

$$R_{\text{source}} = 1000 \ \Omega^3$$
$$R_{load} = 1800 \ \Omega$$
$$C = 2500 \ \text{pF}$$

---

[3]The source and load resistance for this simple filter were not the same. Therefore, the load resistor will not get as much power from the source as it would if it were equal in value. The loss can be calculated as follows. For convenience, assume a 1.0 volt source signal. For a load resistor of 1800Ω, the load voltage will be:

$$V_L = 1.0 \times \frac{1800}{1000 + 1800} = 0.643 \text{ v}$$

The load power will be:

Solution:

The total resistance of this circuit is the parallel combination of the source and load resistances.

$$R_t = \frac{R_s \times R_L}{R_s + R_L}$$

$$= \frac{1000 \times 1800}{1000 + 1800}$$

$$= 642.9 \ \Omega$$

The cutoff frequency is,

$$F_c = \frac{1}{2\pi R_t C}$$

$$= \frac{1}{2\pi \ 642.9 \times 2500 \times 10^{-12}}$$

$$= 99.03 \text{ kHz}$$

$$P_1 = V_L^2/R_L = (0.643)^2/1800 = 0.230 \text{ mW}$$

For a load resistor of 1800Ω, the load voltage will be:

$$V_L = 1.0 \times \frac{1000}{1000 + 1000} = 0.500$$

The load power will be:

$$P_2 \ (0.500)^2/1000 = 0.250 \text{ mW}$$

The loss of power when using an 1800Ω load can be found:

$$\text{Loss (dB)} = 10\log\left(\frac{P_2}{P_1}\right)$$

$$= 10\log\left(\frac{0.250}{0.230}\right)$$

$$= 0.37 \text{ dB}$$

Not very serious for most applications!

For the calculations at 250 kHz we must first calculate the capacitor's reactance.

$$X_c = \frac{1}{2\pi FC}$$

$$= \frac{1}{2\pi\ 250 \times 10^3 \times 2500 \times 10^{-12}}$$

$$= 254.6\ \Omega \text{ (at a } -90° \text{ phase angle)}$$

Now for the attenuation:

$$\text{Attenuation} = 20\log_{10} \frac{X_c}{\sqrt{R_t^2 + X_c^2}}$$

$$= 20\log_{10} \frac{254.6}{\sqrt{642.9^2 + 254.6^2}}$$

$$= 20\log_{10}\ (0.368)$$

$$= 8.68\ \text{dB}$$

And finally phase shift:

$$\text{Phase angle} = \tan^{-1}\ (R_t\ /X_c\ )$$

$$= \tan^{-1}\ (642.9/254.6)$$

$$= \tan^{-1}\ (2.525)$$

$$= 68.4°$$

Now let's check to see how these values compare with those of Figure 3.5. We need the ratio of the actual frequency to the cutoff frequency.

$$\frac{F}{F_c} = \frac{250\ \text{kHz}}{99.03\ \text{kHz}} = 2.52$$

Using this value on the horizontal scale, we see from the graph that our calculations agree with the general single-element response.

3.3 THE PARALLEL-TUNED CIRCUIT

The single *LC* parallel-tuned circuit can be used as a first approximation of the ideal bandpass filter. As with the low-pass filters, the source and load resistance play an important part in the final results. For simplicity, we will start only with a source resistance. The circuit of Figure 3.6 will be used initially. It contains a switch so that either the capacitor or inductor can be switched in either singly or together. The two component values can also be adjusted.

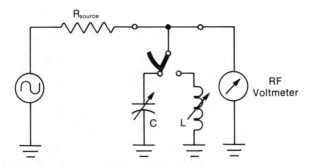

Figure 3.6 Circuit for measuring the frequency response of a parallel-tuned circuit.

For our first situation we will use a small-value capacitor and a large-value inductor. The "capacitor only" response resembles that of a low-pass filter and the "inductor only" response is that of a high-pass filter. The two response curves are shown in Figure 3.7 along with the combined response that would be measured with both the inductor and capacitor in the circuit at the same time.

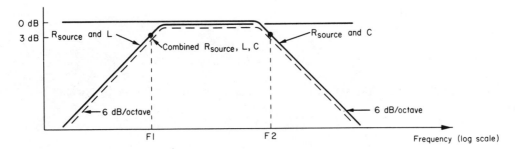

Figure 3.7 Parallel-tuned circuit response with a small capacitor and large inductor.

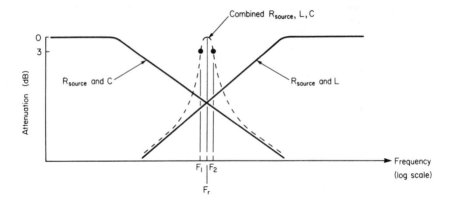

Figure 3.8 Parallel-tuned circuit response with a larger capacitor and a smaller inductor.

As before, the two axes are logarithmic, the overall response curve appears symmetrical, and the skirts end up as straight lines. The slope of each skirt will be 6 dB per octave since only one reactance is significant at any one frequency.[4] For example, on the high-frequency slope, the inductor will have a much higher reactance than the capacitor so, for the parallel combination, the capacitor has the dominant effect.

A different situation results when a larger-value capacitor and a smaller-value inductor are used. The center frequency will remain the same, but the shape of the response curve will change drastically as shown in Figure 3.8.

The two separate curves cross over far down their respective slopes. When both components are connected at the same time, their reactances will be equal and opposite at this crossover frequency and so will cancel each other. The output voltage at this frequency will be the same as the open circuit source voltage and no attenuation occurs. This crossover point is the "resonant frequency" of the tuned circuit. At slightly higher and lower frequencies, the attenuation increases very rapidly, much faster than the slopes shown in Figure 3.7. This fast attenuation change is the result of the two, almost equal reactances, changing in opposite directions.

$$\frac{V_{out}}{V_{source}} = \frac{jX_T}{R + jX_T} \tag{3.11}$$

Where X_T is the total reactance of the capacitor and inductor in parallel.

$$X_T = X_L \parallel X_c$$

---

[4]Even though we have a filter here with two reactive components this is still referred to as a one-pole filter. As has been pointed out, only one reactance is significant at any one frequency.

$$= \frac{-X_L \cdot X_c}{X_L - X_c} \tag{3.12}$$

At lower levels on the attenuation curve, the slopes settle down to the more normal 6 dB per octave. When this happens, we are far enough away from the resonant frequency that only one reactive component is significant; the other will have a very high reactance.

The highest point on the curve is called the *resonant frequency* and is equal to:

$$F_r = \frac{1}{2\pi \sqrt{LC}} \text{ (Hz)} \tag{3.13}$$

The distance between the two 3 dB points (F_1 and F_2) is the "bandwidth" of the filter

$$BW = F_2 - F_1 \text{ (Hz)} \tag{3.14}$$

If the source resistance and the resonant frequency are kept constant and the inductor and capacitor values are changed (in opposite directions to maintain the same resonant frequency), then we end up with a full set of curves as shown in Figures 3.9 and 3.10. One point of interest—on the phase shift curves, the phase shift is always 45° at the 3 dB points of the filter.

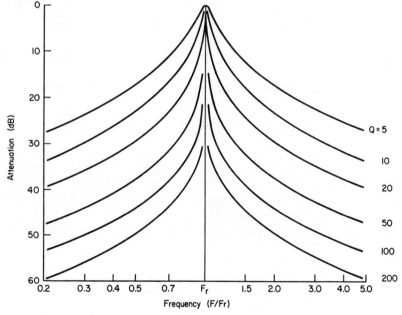

Figure 3.9 Single-pole filter response for loaded Qs from 5 to 200.

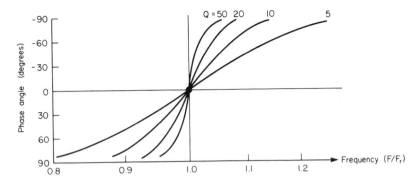

Figure 3.10 Phase shift in a single-pole bandpass filter for loaded Qs from 5 to 50.

A convenient description of the different curves is to use the term *loaded Q*. The higher the loaded Q the narrower the passband, the steeper the initial skirts, and the greater the attenuation before the skirts flatten to the 6 dB slopes.

The loaded Q is calculated from the component reactances at the resonant frequency, assuming that both components are lossless (perfect capacitors and inductors).

$$Q_{loaded} = \frac{R_{source}}{X_L} \text{ or } \frac{R_{source}}{X_c} \qquad (3.15)$$

(R_{source} is the only resistance in the circuit and it is considered in parallel with the tuned circuit).

With the loaded Q we can also calculate the 3 dB bandwith

$$BW_{3dB} = \frac{\text{center frequency}}{\text{loaded } Q} \text{ (Hz)} \qquad (3.16)$$

Now, for any practical application, the tuned circuit will always be used with both a source resistance and a load resistance. How do we convert from our one-resistor calculations to two resistors? The solution is to use an equivalent resistance as shown in Figure 3.11.

From the tuned circuit's point of view, the source and load resistances are in parallel.

$$R_{equivalent} = R_{source} \parallel R_{load}$$
$$= \frac{R_{source} \times R_{load}}{R_{source} + R_{load}} \qquad (3.17)$$

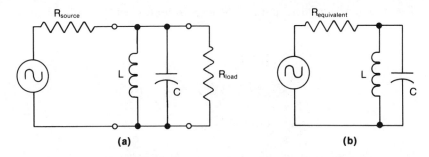

(a)   (b)

Figure 3.11 Thévenin's theorem can be used to convert a circuit with both a source and load resistance (a) into one equivalent resistance (b).

By making the conversion from two resistances down to the one equivalent resistance, we can use the previous equations if we substitute the equivalent resistance for what was the source resistance.

Example 3.2 ══

A tuned circuit, such as that shown in Figure 3.11 (a), has the following component values:

$$R_{source} = 5600\Omega$$

$$R_{load} = 3300\Omega$$

$$L = 150\mu H$$

$$C = .01\mu F$$

Find the circuit's resonant frequency, 3 dB bandwidth, and loaded Q.

Solution:

The resonant frequency is:

$$F_r = \frac{1}{2\pi\sqrt{LC}}$$

$$= \frac{1}{2\pi\sqrt{150\times 10^{-6}\times .01\times 10^{-6}}}$$

$$= 130\ kHz$$

The reactance of the capacitor and inductor will be identical at this frequency

$$X_c = X_L = 2\pi F_r L$$

$$= 2\pi \times 130 \times 10^3 \times 150 \times 10^{-6}$$

$$= 122.5 \; \Omega$$

The equivalent value of the source and load resistances in parallel is,

$$R_{equivalent} = R_{source} \parallel R_{load}$$

$$= \frac{5600 \times 3300}{5600 + 3300}$$

$$= 2076 \; \Omega$$

The loaded Q of the tuned circuit is,

$$Q_{loaded} = \frac{R_{equivalent}}{X_L}$$

$$= \frac{2076}{122.5}$$

$$= 16.9$$

From this and the resonant frequency we can find the bandwidth,

$$BW_{3dB} = F_r / Q_{loaded} = \frac{130 \; kHz}{16.9}$$

$$= 7.69 \; kHz$$

3.4 TWO-ELEMENT, LOW-PASS FILTER

Now we return to low-pass filters to look at a two-element or "two-pole" design. The single-element, low-pass circuit we already looked at suffered one disadvantage—the attenuation slope was extremely slow. This means that signals slightly above the cutoff frequency would pass through the filter with only a small attenuation. It would be nice if we could find a filter that totally eliminated all signals above the cutoff frequency but, unfortunately, such a filter

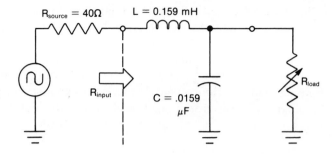

Figure 3.12 Two-element, low-pass filter with a variable load.

hasn't been discovered yet. The two-element filter we are about to look at comes a bit closer to the ideal. We will also see that, in addition to filtering, the circuit may have impedance-changing capabilities and this will be useful in transmitter and receiver circuits.

3.4.1 Sample Filter

A sample filter is shown in Figure 3.12 and some component values are included so that specific examples can be described. The amplitude response for this circuit is shown in Figure 3.13. Different values of load resistance are used and the graph shows that this causes significant changes, especially near the cutoff frequency.

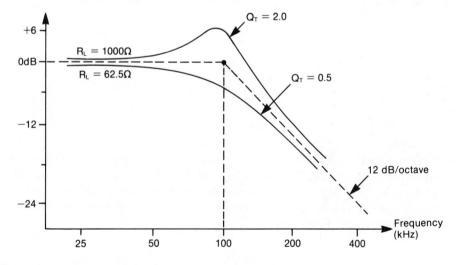

Figure 3.13 Frequency response for the two-element, low-pass filter with different values of load resistance.

All of the curves seem to settle down to a final slope in the stopband of 12 dB per octave, a value typical of any two-element, low-pass filter—6 dB for each component. The 12 dB slope is the result of a voltage divider formed by X_L and X_C where the inductive reactance increases at the same rate as the frequency increases and the capacitive reactance decreases as the frequency increases. In other words, both elements of the voltage divider are changing and in opposite directions. The slope is most pronounced well above the cutoff frequency where X_L is much larger than R_{source} and X_C is much lower than R_{load}.

$$\frac{V_{out}}{V_{source}} \approx \frac{X_C}{X_L + X_C} \text{ (for frequencies well above resonance)}$$

In the immediate vicinity of 100 kHz we see the greatest variation in the curves. This is the series-resonant frequency of the inductor and capacitor and, as we have already seen with the bandpass circuit, resonance effects are very sensitive to resistive loading. In fact, we can still describe the loading in terms of loaded Q but here we have two separate ends to the circuit and so there could be one value of Q at the input and a different value at the output.

$$Q_{input} = \frac{X_L}{R_{source}} \text{ (components in series)}$$

$$Q_{output} = \frac{R_{load}}{X_C} \text{ (components in parallel)}$$

(these Qs are calculated at the series resonant frequency of L and C)

The final shape of the curve depends on the combination of these two values.

$$Q_{total} = \frac{Q_{input} \times Q_{output}}{Q_{input} + Q_{output}}$$

A total Q greater than 0.5 will produce a peak in the filter and a total Q less than 0.5 will never have a peak. The flattest response occurs exactly at the value 0.5. Normally, a designer would try and make the input and output Q have the same value because this would allow the maximum amount of power pass from the source through to the load.

$$Q_{input} = Q_{output} \text{ (for maximum power transfer)}$$

3.4.2 Impedance Transformation

When the total Q is higher than 0.5, the peaks in the response curves produce a higher output voltage than the input voltage. If this seems strange, remember that a step-up transformer does the same thing. In both cases, the power entering the filter (or transformer) from the source will be the same as that going to the load. In other words, a higher output voltage must be accompanied by a lower output current. This automatically means that the load resistor must be a higher value than the input resistance of the network. Therefore, the two-element filter is acting as an impedance-changing circuit in the same way that the transformer does. The one big difference is that the filter only provides this impedance change at the peak frequency and so is, therefore, a very narrow band circuit. We will see these circuits used in the chapter on transmitters.

3.4.3 Signal Delay

In Chapter 1 we mentioned the delay distortion that could occur as complex signals pass through amplifiers and filters. For our two-element filters, the time it takes for sine waves of various frequencies to pass through is shown in Figure 3.14. Only one value of load resistor provides a constant time delay at all frequencies. If this filter were to be used for a television signal, the designer would have to be very careful that the delay response stayed flat otherwise the picture on the TV screen could be slightly smeared. Digital signals could also suffer from passing through filters that have variations in their time delay. The result

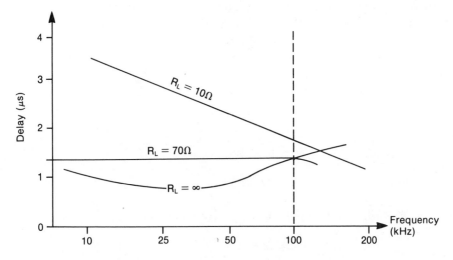

Figure 3.14   Sine wave delay times for a two-element, low-pass filter. The value of load resistor is important.

would be *intersymbol interference* which causes errors when the computer tries to read the digital signals. This will be discussed more in the data communications chapter.

Example 3.3

For the circuit shown below, sketch the approximate frequency response. Use the curves of Figure 3.13 to get a general idea of what the result will be.

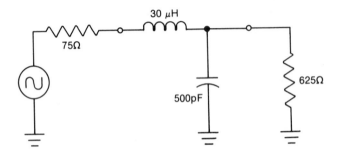

Figure 3.15

Solution:

The shape of the amplitude versus frequency curve is governed by the total Q of the circuit. To determine this, we first find the resonant frequency of L and C and then the Q of each end of the filter.

The LC resonant frequency is:

$$F_r = \frac{1}{2\pi\sqrt{LC}}$$

$$= \frac{1}{2\pi\sqrt{30 \times 10^{-6} \times 500 \times 10^{-12}}}$$

$$= 1.30 \text{ MHz}$$

The reactance of each component at this frequency is:

$$X_C = X_L = 2\pi F_r L$$

$$= 2\pi 1.3 \times 10^6 \times 30 \times 10^{-6}$$

$$= 245 \ \Omega$$

The input and output values of the loaded Q are:

$$Q_{input} = \frac{X_L}{R_{source}}$$

$$= \frac{245}{75}$$

$$= 3.27$$

$$Q_{output} = \frac{R_{load}}{X_c}$$

$$= \frac{625}{245}$$

$$= 2.55$$

These combine to form a total Q of:

$$Q_{total} = \frac{Q_{in} \times Q_{out}}{Q_{in} + Q_{out}}$$

$$= \frac{3.27 \times 2.55}{3.27 + 2.55}$$

$$= 1.43$$

Using this value, we can look at Figure 3.13 to find the general shape of the curve. A rough sketch of the response is shown in Figure 3.16. A more accurate curve would require more involved mathematics.

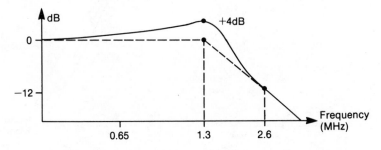

Figure 3.16

3.5 THREE-ELEMENT FILTERS

Having looked at two-element filters and finding that they come a bit closer to the ideal, let us take a quick look at the three-element, low-pass filter. A general circuit is shown in Figure 3.17.

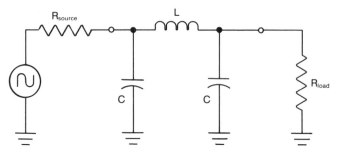

Figure 3.17 Three-element, low-pass filter.

If the two capacitors have the same value and the source and load resistances are the same, the circuit is a symmetrical filter and it can have frequency responses as shown in Figure 3.18. Again, the total Q of the filter sets the overall shape of the curve. For a symmetrical three-element filter:

$$Q_{input} = Q_{output} = \frac{R_{load}}{X_c}$$

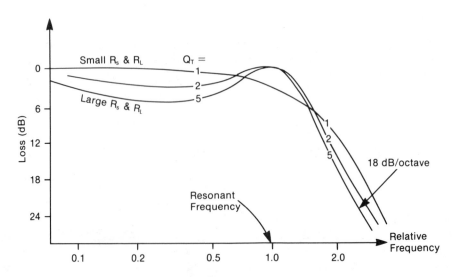

Figure 3.18 Frequency response of a symmetrical three-element, low-pass filter having various values of loaded Q.

(The reactance is calculated at the resonant frequency of L and $C/2$.)
The total Q is then:

$$Q_{total} = \frac{Q_{input}}{2} = \frac{Q_{output}}{2}$$

The final slope of the filter settles down to 18 dB per octave. Again, this is a slope of 6 dB for each reactive component in the filter. The higher Q curves have peaks and their initial slopes at the beginnings of the stopband are steeper than the 18 dB value. The lower Q curves have no peaks and their initial slopes in the stopband are less than the final 18 dB value. The higher Q and the steeper initial slopes can be used to make a more efficient filter because, for a given cutoff frequency, they will provide more attenuation in the stopband. However, as with most benefits, there are problems. Higher Q filters have delay responses that are not constant and this could cause serious delay distortion in some applications. The curves of Figure 3.19 show how the delay changes with the loaded Q of the three-element filters.

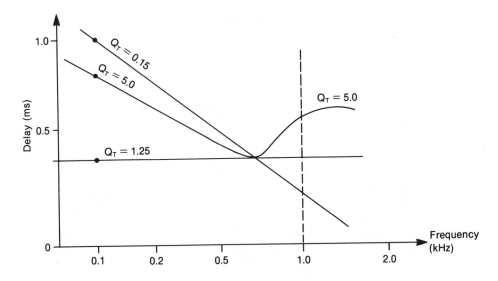

Figure 3.19 Delay times for sine waves to pass through three-element filters that have various values of loaded Q.

3.6 PRACTICAL-TUNED CIRCUITS

In Chapter 2 we studied the flaws that exist in practical capacitors and inductors. Now we will go back to the simple bandpass filter discussed in this chapter and see how these practical components change their characteristics and how the effects can be calculated.

The most significant problem concerns the bandwidth of the final filter. Bandwidth of the filter is inversely proportional to the loaded Q of the circuit and the problem is that the loaded Q can never be greater than the Q of the individual components, with the inductor being the biggest problem. This means that it will be very difficult to build very narrow bandpass filters since high Q inductors would be needed but are not readily available.

A second problem is that the component losses also use up some of the energy that should pass through the filter and so a small loss occurs even at the peak frequency.

To study these and related problems, consider the tuned circuit of Figure 3.20 in which both the capacitor and the inductor have losses that are being represented as series resistances. The inductor loss has been chosen to be greater than the capacitor loss since this is what normally happens in a practical situation.

The presence of the "loss" resistors in series with each component will cause several problems. One is the lowering of the overall loaded Q and a corresponding widening of the 3 dB bandwidth. Another is the increased loss at the peak frequency. A third, and usually minor problem, is a slight shift of the resonant frequency caused by the unequal losses in the two components. If each component has a moderate Q (10 or more), this shift can often be ignored.

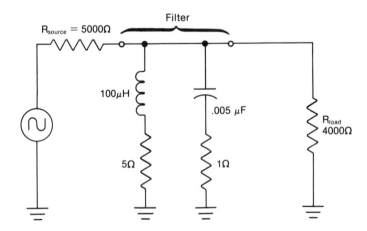

Figure 3.20 A single-tuned circuit with losses in both the capacitor and the inductor.

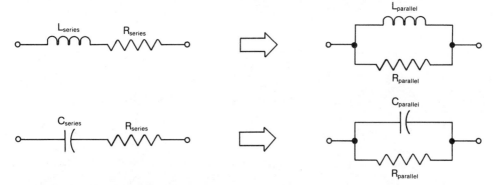

Figure 3.21 Series-to-parallel conversion. The equivalent values are accurate only for the one frequency for which the conversion is made.

3.6.1 Series-to-Parallel Conversion

To simplify the calculations on this circuit, it is convenient to make a conversion from a resistance in series with each component to one in parallel. This conversion is only accurate at the one frequency but is still very useful. Figure 3.21 shows the conversion.

The conversion is made so that the impedance of the parallel components will be exactly the same as that of the original series components. This means that the Q of the series circuit will have to be the same as that of the parallel circuit.

If we know the Q for each of the series circuits:

$$Q = \frac{X_{Lseries}}{R_{series}} \ or \ \frac{X_{c \ series}}{R_{series}}$$

Then the values of the parallel components are:

$$X_{parallel} = \left(1 + \frac{1}{Q^2} \right) \cdot X_{series}$$

$$R_{parallel} = (1 + Q^2) \cdot R_{series}$$

If the Qs of the circuits are higher than 10, the equations can be simplified (with less than 1% error) to:

$$X_{parallel} \approx X_{series}$$

$$R_{parallel} \approx Q^2 \cdot R_{series}$$

Now we are ready to use this conversion on the tuned circuit of Figure 3.20. Our goal is to be able to sketch the frequency response of that circuit.

First we will find the resonant frequency based only on the values of the two reactive components.

$$F_r = \frac{1}{2\pi\sqrt{LC}} = \frac{1}{2\pi\sqrt{100 \times 10^{-6} \times .005 \times 10^{-6}}}$$

$$= 225 \cdot 08 \text{ kHz}$$

The reactances of the two components will be equal at this frequency.

$$X_c = X_L = 2\pi F_r L$$

$$= 2\pi \times 225.08 \times 10^3 \times 100 \times 10^{-6}$$

$$= 141.42 \ \Omega$$

The Q of the inductor and its series resistor is:

$$Q_1 = \frac{X_L}{R_S} = \frac{141.42}{5} = 28.284$$

The Q of the capacitor and its series resistor is:

$$Q_2 = \frac{X_C}{R_S} = \frac{141.41}{1} = 141.42$$

Now we can make the conversion to the parallel equivalents. Even though we could get away with the more approximate formulas since both Qs are greater than 10, we will stick with the exact ones.

$$X_{L\ parallel} = \left(1 + \frac{1}{Q_1^2}\right) \cdot X_{L\ series}$$

$$= \left(1 + \frac{1}{28.284^2}\right) \times 141.42$$

$$= 141.597 \ \Omega$$

$$X_{C\ parallel} = \left(1 + \frac{1}{Q_2^2}\right) \cdot X_{C\ series}$$

$$= \left(1 + \frac{1}{141.42^2}\right) \times 141.42$$

$$= 141.43$$

The two equivalent parallel reactive components are very slightly different which indicates that we are not exactly at the true resonant frequency. The correct point is 135 Hz lower than we calculated because of the unequal Qs of the original inductor and capacitor. This difference is normally insignificant.

The parallel resistor values can also be calculated. We will need these to determine the total resistive loading and, therefore, the total loaded Q of the tuned circuit. For the inductor losses:

$$R_{parallel} = (1 + Q_1^2) \cdot R_{series}$$

$$= (1 + 28.28^2) \times 5$$

$$= 4005 \ \Omega$$

For the capacitor losses:

$$R_{parallel} = (1 + Q_2^2) \cdot R_{series}$$

$$= (1 + 141.42^2) \times 1$$

$$= 20.00 \ k\Omega$$

The complete equivalent circuit now appears in Figure 3.22. Remember that it is only exactly correct at the calculated resonant frequency. However, we will stretch the truth a bit and assume that it is reasonably correct over the 3 dB passband.

At resonance, the inductive reactance and the capacitive reactance cancel each other and the only part of the tuned circuit remaining is the two parallel resistors. Together, the two resistors have a value of 3337 ohms, and this will be referred to as the *combined loss resistance*.

The tuned circuit now has three resistors in parallel with it—the source, the load, and the combined loss resistance. The parallel combination is:

$$R_T = \cfrac{1}{\cfrac{1}{R_{source}} + \cfrac{1}{R_{loss}} + \cfrac{1}{R_{load}}}$$

$$= \cfrac{1}{\cfrac{1}{5k} + \cfrac{1}{3.337k} + \cfrac{1}{4k}} = 1334\Omega$$

From this total resistance, we can find the final loaded Q:

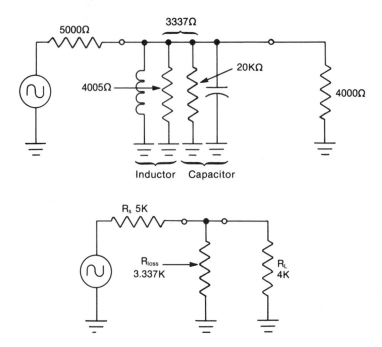

Figure 3.22 Equivalent tuned circuit with component losses shown as parallel resistors.

$$Q_{loaded} = R_T / X_L$$

$$= \frac{1334}{141.6} = 9.421$$

The 3 dB bandwidth would then be:

$$BW_{3dB} = \frac{F_r}{Q_{loaded}}$$

$$= \frac{225.08 \text{ kHz}}{9.421}$$

$$= 23.90 \text{ kHz}$$

3.6.2 Insertion Loss

In addition to the slight shift in resonant frequency and the lowering of the total loaded Q, the component losses will also prevent some of the source signal from reaching the load even at the peak frequency. This is called *insertion loss*. It is the loss in load voltage caused by inserting the filter, with its losses, between the source and the load. This loss can be calculated by comparing the voltage across the load, both with the parallel loss resistors and without them. Without the filter:

$$\frac{V_{\text{load}}}{V_{\text{source}}} = \frac{4\text{ k}}{5\text{ k} + 5\text{ k}} = 0.444$$

With the filter and its losses:

$$\frac{V_{\text{load}}}{V_{\text{source}}} = \frac{4\text{ k} \parallel 3.337\text{ k}}{5\text{ k} + (4\text{ k} \parallel 3.337\text{ k})}$$

$$= 0.2668$$

The insertion loss is then:

$$\text{IL} = 20\log_{10}\left(\frac{0.2668}{0.444}\right)$$

$$= 4.42\text{ dB}$$

The frequency response of the complete tuned circuit is shown in Figure 3.23. Notice that at the extreme ends of the skirts, the attenuation stops at some

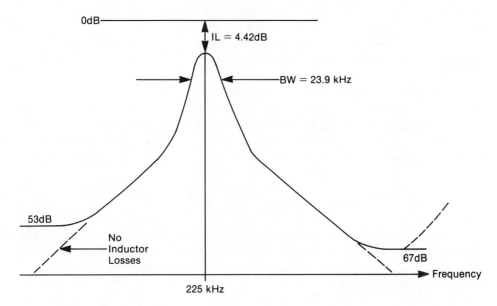

Figure 3.23 Frequency response of a tuned circuit with loss components as shown in Figure 3.20.

value. This is another problem caused by the component losses. At low frequencies, the impedance of the inductor cannot go below 5 Ω and at high frequencies, the impedance of the capacitor cannot go below 1 Ω. These constant impedances produce the flat skirt portions.

In practice, the tuned circuit could be even more complex. The loss resistances could change with frequency and each component could, by itself, have a separate resonant frequency. The capacitor is usually the most likely offender if it is connected into a circuit with excessively long leads. On the upper skirt then, the attenuation would start to rise again as the capacitor actually becomes inductive above its self-resonant frequency.

Designing tuned circuits requires careful attention to the components used.

3.7 IMPEDANCE MATCHING

For many applications of tuned circuits, the source and load resistances are not conveniently close to the same value. A common example is in a radio receiver where a signal is picked up from a 50Ω antenna (the source) and is to be amplified by a transistor that has an input resistance of 800Ω (the load). Without any impedance matching, the transistor will not absorb as much of the available power from the antenna as it could if properly matched; in other words, some potential gain is being wasted (6.55 dB). For the radio receiver, a tuned circuit is usually placed between the antenna and the first amplifier to get rid of interfering signals. It would be nice if this circuit could also improve the impedance match between the antenna and the transistor. Three methods of providing matching along with the filtering are shown in Figure 3.24.

All three circuits produce the same end result, matching a lower value source resistance to a higher value load resistance, while retaining the frequency response of a single-tuned circuit. The circuit of Figure 3.24 (a) uses a tapped inductor as a transformer to step up the source to the equivalent value:

$$R'_{source} = R_{source} \times \left(\frac{n_2}{n_1} \right)^2 \text{ for } Q_{loaded} > 10$$

The circuit of Figure 3.24 (b) operates in exactly the same way but has no DC connection between input and output which might be handy in some cases. The circuit of Figure 3.24 (c) uses a tapped capacitor arrangement to step up the source resistance:

$$R'_{source} = R_{source} \times \left(1 + \frac{C_1}{C_2} \right)^2 \text{ for } Q_{loaded} > 10$$

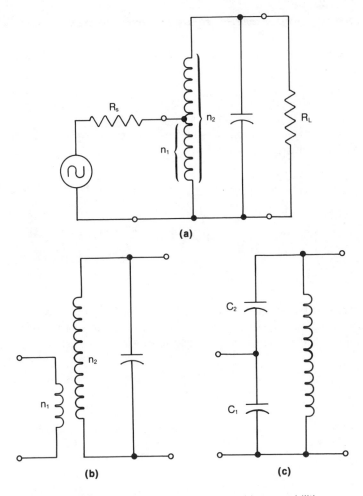

Figure 3.24 Single-tuned circuits with impedance-matching capabilities.

The two capacitors of this circuit act in series when the resonant frequency is being calculated. The equivalent capacitance is;

$$C_{\text{equivalent}} = \frac{C_1 \times C_2}{C_1 + C_2}$$

To illustrate the use of these techniques, let us try the following example.

Example 3.5 ═══════════════════════════════════

Calculate the center frequency and bandwidth of the tuned circuit shown in Figure 3.25. The capacitor is perfect but the inductor itself has a Q of 100.

Solution:

The 50Ω resistance of the antenna is stepped up to an equivalent value that is placed across the entire tuned circuit.

$$R'_{source} = R_{source} \times \left(1 + \frac{C_1}{C_2}\right)^2$$

$$= 50 \times \left(1 + \frac{3000}{1500}\right)^2$$

$$= 450 \ \Omega$$

which happens to be the same value as the transistor's input resistance. Maximum power transfer will occur (except for some losses in the coil).
   The equivalent value of the two capacitors in series is:

$$C_{equivalent} = \frac{C_1 \times C_2}{C_1 + C_2}$$

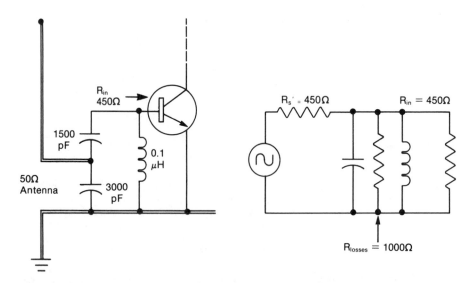

Figure 3.25 The input circuit of a radio receiver (a) and its equivalent circuit (b).

$$= \frac{3000 \times 1500}{3000 + 1500}$$

$$= 1000 \text{ pF}$$

The resonant frequency of this equivalent capacitor and the inductor is:

$$F_r = \frac{1}{2\pi \sqrt{LC_{\text{equivalent}}}},$$

$$= \frac{1}{2\pi \sqrt{0.1 \times 10^{-6} \times 1000 \times 10^{-12}}}$$

$$= 15.92 \text{ MHz}$$

Our next step is to find how serious the losses in the inductor are. First we need the reactance of the inductor:

$$X_L = 2\pi F_r L$$

$$= 2\pi \times 15.915 \times 10^6 \times 0.1 \times 10^{-6}$$

$$= 10.0 \ \Omega$$

Then we can find the equivalent parallel loss resistance:

$$R_{\text{parallel}} = Q \times X_L$$

$$= 100 \times 10.0$$

$$= 1000 \ \Omega$$

The complete equivalent circuit is shown in Figure 3.25 (b). The total resistive load on the tuned circuit is:

$$450 \ \Omega \ \| \ 1000 \ \Omega \ \| \ 450 \ \Omega \ = \ 183.7 \ \Omega$$

The loaded Q of the complete circuit is then:

$$Q_{\text{loaded}} = \frac{R_{\text{total}}}{X_L}$$

$$= \frac{183.7}{10}$$

$$= 18.37$$

And, finally, we can calculate the bandwidth of the circuit:

$$BW_{3dB} = \frac{F_r}{Q_{loaded}}$$

$$= \frac{15.92 \times 10^6}{18.37}$$

$$= 866.3 \text{ kHz}$$

3.8 COUPLED FILTERS

The single-tuned circuit we have been discussing was our first approximation to the ideal, flat-topped, straight-sided bandpass filter. But it isn't good enough for many applications; the passband is quite rounded and the skirts are not steep enough and even more, the two interact. If we want steeper skirts, the bandwidth must shrink and if we want wider bandwidths, the skirts must rise (refer back to Figure 3.9). By coupling a number of single-tuned circuits together, a much closer approximation of the ideal filter is obtained and bandwidth and skirt characteristics become a bit more independent.

3.8.1 Transistor Coupling

There are many coupling possibilities, each with its own characteristics. The simplest is to put transistors in between individual tuned circuits as shown in Figure 3.26.

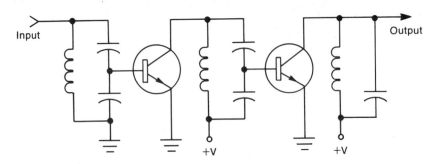

Figure 3.26 One-way coupling of single-pole filters.

The transistor is assumed to be a one-way device which means that each tuned circuit is completely isolated from the others and is loaded only by constant input and output impedances of transistors. In practice, this isn't exactly true as we will see later, but transistor coupling does work reasonably well.

If all the tuned circuits are similar, have the same loaded Q, and are set to the same frequency, the bandwidth will be reduced, depending on the number of stages. This technique is called *synchronous tuning*. While it is somewhat useful if narrow bandwidths are required, the skirts still resemble those of the high Q tuned circuits and the passband is still rounded.

A different result is obtained if each tuned circuit is set to a different frequency; this is called *stagger tuning*. The overall bandwidth will now be wider but the passband can be very flat and the skirts steeper, producing a better shape factor. As with synchronous tuning, the overall response is simply the sum of the individual responses in decibels. By selecting the individual center frequencies and loaded Qs, virtually any shape can be produced. The ultimate restriction is that for narrow bandwidths, the Qs of the individual circuits become too high for practical LC components. This is where the higher Q of the mechanical, crystal, and ceramic resonators are required.

Three resonant circuits that are stagger tuned to provide a wider, flatter response are shown in Figure 3.27. Notice that the combined result shows considerable attenuation. This happens because only one peak touches the 0 dB reference line at a time while the other two are producing some attenuation. The combined result must be the sum of all three attenuations at each fre-

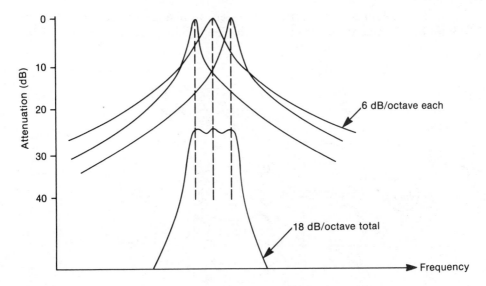

Figure 3.27 Stagger tuning of three resonant circuits to produce a wider, flatter bandpass and steeper skirts.

quency. In actual use, the coupling transistors will provide some gain and this would be added to the overall results.

Stagger tuning is used for many radar and television intermediate frequency (IF) amplifiers.

3.8.2 Inductive Coupling

The transistor coupling we have just discussed is essentially one way. The different tuned circuits do not interact and the final result is simply the sum of the individual circuits. Inductive coupling is considerably more complex.

One way to produce this two-way coupling is to position two coils physically close together so that part of their magnetic fields overlap as shown in Figure 3.28. This is a common commercial product used for IF transformers as shown in Figure 3.29. Because of the two-way magnetic coupling, the secondary impedance will affect the primary voltages and currents, and the primary impedance will affect those of the secondary. How drastic the effect will be depends, in part, on how much of the magnetic fields overlap or couple.

The degree of coupling is indicated by k. If only 10% of the primary's field is picked up by the secondary then $k = 0.1$. If all of the field is picked up by the secondary then $k = 1.0$, a condition that is most desirable for 60 Hz power transformers and wideband audio transformers. However, for the majority of the double-tuned, narrowband IF transformers, values of k less than 0.1 are very common.

For calculation purposes, it is convenient to change the circuit of Figure 3.28 into the equivalent one of Figure 3.30. The inductances shown are individ-

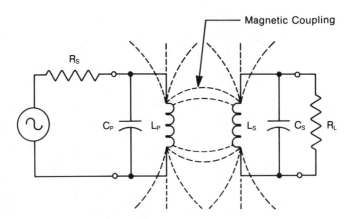

Figure 3.28 Inductive coupling of two identical circuits occurs when magnetic fields overlap.

Figure 3.29 Commercial double-tuned IF transformers.

ual components and no coupling occurs between them. The one inductance, labeled M, is common to both tuned circuits and it represents the coupling element. It is called the *mutual inductance* to indicate its common nature. The primary and secondary inductances are made identical here to further simplify the circuit.

$$M = k\sqrt{L_p L_s}$$

$$= kL \text{ (if } L_p = L_s)$$

where k = degree of coupling
L_p = primary inductance (H)
L_s = secondary inductance (H)

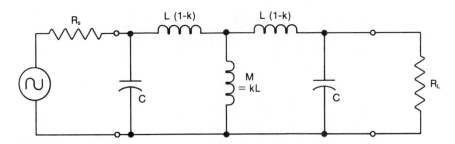

Figure 3.30 Equivalent circuit of the double-tuned transformer with a mutual inductance to represent magnetic coupling.

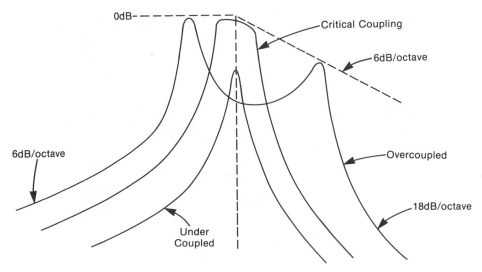

Figure 3.31 Frequency response of a double-tuned circuit with various degrees of magnetic coupling.

Where the primary and secondary coils are far apart, k is small so the coupling inductance is small. The other two inductances of the equivalent circuit will be almost as large as the original primary and secondary inductances. The overall characteristics then strongly resemble the sum of the two individual circuits. As the coils are moved closer together, the value of k increases, the coupling inductance (kL) increases, and the other two inductances, $(1-k)L$, become correspondingly smaller. The overall characteristics of the circuit change with coil spacing are shown in Figure 3.31.

When the coils are spaced far apart and very little coupling occurs, a single peak appears at the same frequency as that of the individual tuned circuits. As the coils move closer together, the height of this peak increases and its bandwidth increases. Finally, a point is reached where maximum energy is being transferred from primary to secondary and the peak is, therefore, as high as it can possibly be. The corresponding value of k at this point is called *critical coupling* (k_c). Its numeric value depends on the loaded Qs of the primary and secondary circuits.

$$k_c = \frac{1}{\sqrt{Q_p\, Q_s}}$$

For closer spacing, the secondary voltage curve starts to show a double-peaked response with the higher frequency peak at a slightly lower amplitude than the lower peak. As coupling increases, the two peaks move away from each other and the dip in between becomes deeper.

When the two tuned circuits are overcoupled, the low frequency peak slowly moves toward a frequency that is 0.707 times the individual resonant frequency. This happens because, at total coupling ($k = 1.0$) the two inductors blend into one that has the same total inductance as each original coil. (There will only be one magnetic field instead of the original two.)

The high frequency peak moves faster. As the coupling increases, this peak moves toward infinity and becomes smaller in size all the time. When $k = 1.0$, the peak reaches infinity but its amplitude will have been reduced to zero and therefore won't exist. This is correct because the complete circuit has been reduced to one tuned circuit that can only have one resonant frequency.

The most useful range of coupling is in the immediate vicinity of the critical value. Slight undercoupling provides more linear phase response (constant delay) across the passband but the amplitude versus frequency response is rounded. This is a desirable condition for the IF amplifiers of FM receivers where phase distortion is more serious than amplitude distortion. For AM receivers, slight overcoupling is used, resulting in a flatter amplitude response but a more nonlinear phase response.

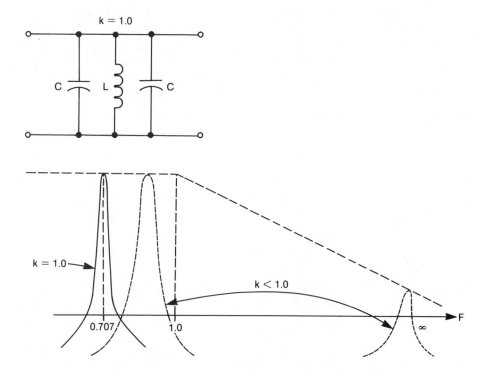

Figure 3.32 As total coupling ($k = 1.0$) is approached, the upper peak moves toward an infinite frequency and zero amplitude.

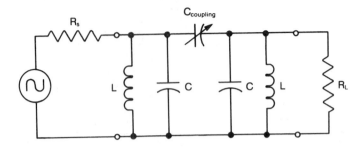

Figure 3.33 Capacitive coupling of two resonant circuits.

3.8.3 Capacitor Coupling

Using a capacitor to couple two resonant circuits together, as shown in Figure 3.33, produces the same benefits as the inductive coupling: flat passband and steep skirts, although the frequency response is almost a complete mirror image of the inductively coupled circuit.

The frequency response again shows a double peak if overcoupled, but this time the lower amplitude peak is on the lower frequency side. It still becomes smaller at the rate of 6 dB per octave as the coupling increases. The upper peak remains fixed, unlike the inductive coupling case. This difference occurs because one of the components of the circuit changes—the coupling capacitor—while all the other values remain fixed. The general frequency response curve is shown in Figure 3.34.

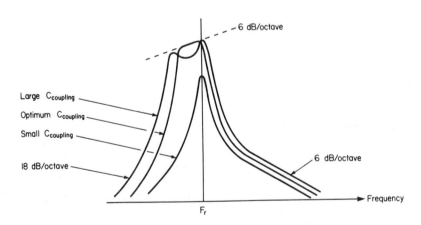

Figure 3.34 By changing the size of the coupling capacitor, the bandwidth and the flatness of the passband can be altered. A small coupling capacitor produces an undercoupled response and a large value produces an overcoupled response.

At critical coupling, the frequency response has a flat passband and skirts that are initially very steep. The passband has a slope of 6 dB per octave, but for most narrow band applications the bandwidth is a small fraction of an octave and so the tilt is completely negligible. The skirts on the high frequency side eventually settle down to a gentle 6 dB per octave slope. The low-frequency skirts are a steeper 18 dB per octave. Again, when narrow bandwidths are used, the attenuation values are quite large by the time these skirts begin to flare out.

The value of the coupling capacitor for critical coupling is:

$$C_{coupling} = \frac{C}{Q_{loaded}}$$

where C is the resonating capacitor of the parallel circuit.

As with the inductively coupled circuit, the final 3 dB bandwidth will be approximately $\sqrt{2}$ times the bandwidth of each of the tuned circuits (critical coupling only).

Example 3.6 ━━━━━━━━━━━━━━━━━━━━━━━━━━━━━━━━━

Design a capacitively critically coupled double-tuned circuit for a 3 dB bandwidth of 20 kHz at a center frequency of 455 kHz. The source and load resistors will each be 2200 Ω.

Solution:

First select a resonant frequency for the individual tuned circuits. From Figure 3.34, we can estimate that the resonant frequency is about halfway between the center frequency and the upper 3dB point. For our example, this would be approximately 460 kHz.

The final bandwidth must be 20 kHz so the bandwidth of each individual tuned circuit will be:

$$BW_{single} = \frac{20 \text{ kHz}}{\sqrt{2}}$$

$$= 14.14 \text{ kHz}$$

The loaded Q of each circuit will be:

$$Q = \frac{F_r}{BW}$$

$$= \frac{460 \text{ kHz}}{14.14 \text{ kHz}}$$

$$= 32.53$$

Now we can calculate the capacitive and inductive reactances:

$$X_L = X_C = \frac{R_{source}}{Q}$$

$$= \frac{2200}{32.53}$$

$$= 67.63$$

These are the reactances at the 460 kHz resonant frequency. The component values are:

$$L = \frac{X_L}{2\pi F}$$

$$= \frac{67.63}{2\pi \times 460 \times 10^3}$$

$$= 23.399 \mu H$$

and
$$C = \frac{1}{2\pi F X_c}$$

$$= \frac{1}{2\pi \times 460 \times 10^3 \times 67.63}$$

$$= 5116 \text{ pF}$$

And finally, the coupling capacitor value is:

$$C_{coupling} = \frac{C}{Q_{loaded}}$$

$$= \frac{5116}{32.53}$$

$$= 157.3 \text{ pF}$$

The circuit diagram for the final filter is shown in Figure 3.35.

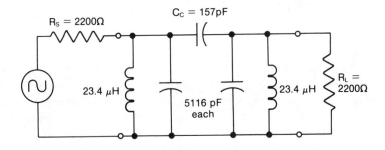

Figure 3.35   455 kHz double-tuned filter with a 20 kHz bandwidth.

3.9   NARROW BANDPASS FILTERS

In many receiver circuits, narrow bandpass filters are required with very steep skirts. This is accomplished by using larger numbers of high Q-tuned circuits and taking advantage of their initial fast attenuation. Any of the previous techniques can be extended to build filters with as many as eight, or even more, tuned circuits. However, it is soon discovered that the Qs of the inductors that are required just aren't practical; remember that the loaded Q can never be higher than the Q of the individual components. Even with Litz wire, Qs above 200 aren't very common.

The solution is to look for different kinds of tuned circuits using methods other than inductors and capacitors. Four common techniques use ceramic, quartz, saw, or mechanical resonators; all have Q's above 200, with some quartz resonators exceeding 10,000. Each technique tends to have its own range of desirable center frequencies and bandwidths, although some overlap is found as shown in Figure 3.36.

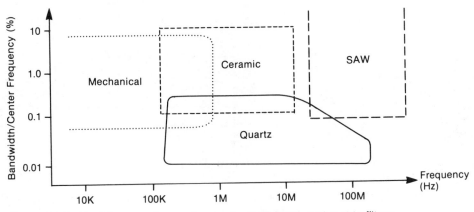

Figure 3.36   Useful range of ceramic, mechanical, SAW, and quartz filters.

3.9.1 Quartz Crystal Resonators

Thin slices cut from crystals of pure quartz (SiO_2) have two interesting properties. First, they are piezoelectric. This means that, with wires attached, a bending of the quartz slice will produce a voltage and, conversely, if a voltage is
applied, the slice will bend. A second property of the quartz slices is that they
have a distinct preference to mechanically vibrate at one frequency and so act
as very high Q resonators. This frequency is determined by the physical dimensions of the slice and tends to be extremely stable. Quartz crystal resonators,
therefore, appear to act as very high Q ($> 10{,}000$) and highly stable electrical
tuned circuits when measured between their two electrodes. The resonant frequency will change only one cycle out of every 10^9 in one day.

Figure 3.37 shows a typical grown crystal from which the slices are cut.
The plane of the cut has a great effect on the temperature stability of the resonator's frequency as shown in Figure 3.38. A change of less than $0.3°$ in the
angle of cut can make a large difference in the frequency change if the temperature varies. The temperature change can be minimized if the crystal is enclosed in an "oven" that keeps the temperature constant at some higher
temperature—preferably at one where the curve runs horizontal as indicated by
the dot in Figure 3.38.

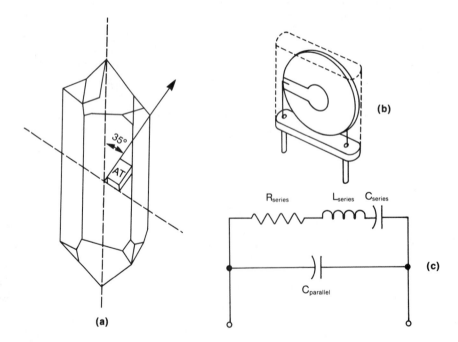

Figure 3.37 Crystal of quartz showing standard "AT" cutting plane (a). A round slice
mounted in a holder (b) and (c) its electrical equivalent circuit.

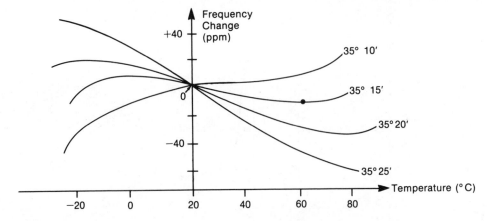

Figure 3.38 Temperature changes will shift the resonant frequency of AT-cut quartz resonators. The amount depends on the cutting angle from the original crystal.

The equivalent circuit of Figure 3.37 (c) shows that the mechanical proper-ties and the electrodes form a combination series-parallel resonant circuit. The variation of the resonator's impedance with frequency is shown in Figure 3.39. The horizontal frequency scale is highly magnified here showing that the series and parallel resonant frequencies occur very close together. This graph is typi-cal of a 10.7 MHz "crystal" with the following physical and electrical characteristics:

$$\text{thickness} = 0.16\text{mm}$$

$$
\begin{aligned}
R_{series} &= 34\Omega \\
L_{series} &= 23\text{mH} \\
C_{series} &= .008\text{pF} \\
C_{parallel} &= 2\text{ pF}
\end{aligned}
$$

Remember that the electrical values of R, L, and C are simply analogs of the resonator's mechanical properties—you won't find a coil of wire inside a crystal can.

The huge equivalent inductance and the relatively low series resistance result in a very high Q. At the 10.7 MHz resonant frequency:

$$Q = \frac{X_L}{R_s} = \frac{2\pi \times 10.7 \times 10^6 \times 23 \times 10^{-3}}{34}$$

$$= 45{,}480$$

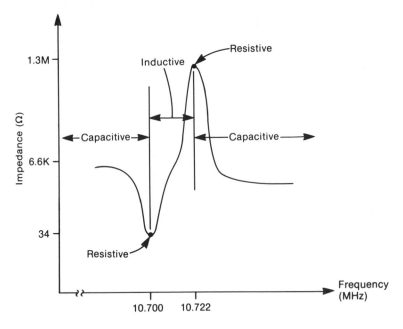

Figure 3.39 Impedance variation of a 10.7 MHz quartz crystal with frequency.

The very high Q and the extreme stability of the resonant frequency make the quartz "crystal" resonator very useful for oscillator circuits (Chapter 4) and for narrow bandwidth filters.

3.9.2 Crystal Filters

If a quartz crystal is placed between a signal source and a load it will act as a narrow bandpass filter at its series-resonant frequency. This is shown in Figure 3.40.

Using two crystals resonant at slightly different frequencies, it is possible to make a bridge or lattice type of filter. At the two series-resonant frequencies, the output signal is large but, at slightly higher and lower frequencies, the bridge will be balanced (i.e., have equal reactances in both arms) and so have no output. Very steep skirts are the result.

Several of these lattice sections can be added in series (cascaded) to form very elaborate filters with good shape factors. Four-, six-, and eight- pole crystal filters are quite common. The insides of an eight-pole commercial design are shown in Figure 3.42.

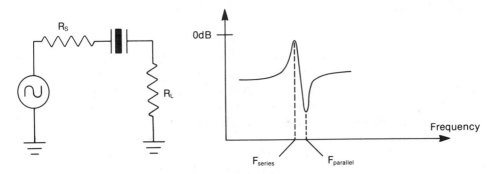

Figure 3.40 Simple crystal filter and its frequency response.

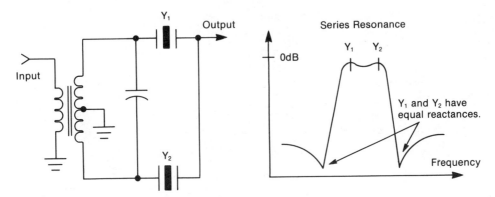

Figure 3.41 Lattice crystal filters use pairs of crystals to obtain very steep skirts.

Figure 3.42 Interior view of an eight-pole crystal filter.

3.9.3 Ceramic Filters

As with quartz, certain types of ceramic materials have piezoelectric properties. The difference is that the temperature stability is not quite as good and the Q is lower. For these reasons, ceramic resonators are rarely used in oscillator circuits but are often used for low-cost, good shape factor bandpass filters where very narrow bandwidths are not required.

Although individual ceramic resonators can be used in the same manner as crystal resonators, it is more economical to plate several electrodes on one disk and have it act as a pair of *coupled resonators*. (This technique is also used with quartz crystals to build *monolithic crystal filters*.)

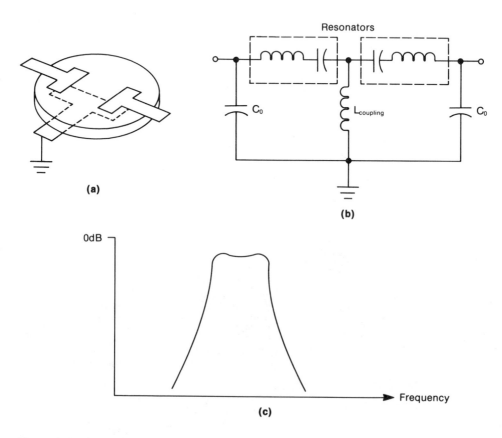

Figure 3.43 Two-pole coupled ceramic resonator (a), its equivalent circuit (b) and (c) frequency response.

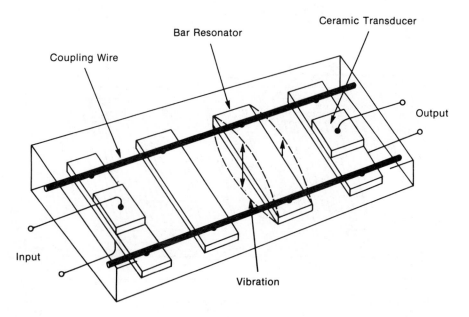

Figure 3.44   Four-resonator mechanical filter.

3.9.4   Mechanical Filters

Any piece of metal will have a natural resonant frequency and, if the right material and shapes are used, the Q can be very high (over 10,000) and the temperature stability very good (one cycle per million). The one problem is that metal is not piezoelectric; applying a voltage will simply result in sparks from a short circuit. Mechanical filters, therefore, require added transducers that convert the electrical oscillations into mechanical ones and then back again. The transducers can be piezoelectric ceramic or magnetostrictive ferrites. The resonators can be any convenient mechanical shape, usually bars or discs, and they are coupled together by thin wires. A four-resonator filter with ceramic input and output transducers is shown in Figure 3.44.

3.9.5   Surface-Acoustic-Wave Filters

In solid materials, acoustic waves travel at the speed of sound, a speed that is much slower than that of radio waves. This means that a wavelength for the acoustical signal is much shorter than an electrical wavelength and so periodic structures can be built in very small areas. The result is a filter device in which the amplitude and phase response can be independently adjusted.

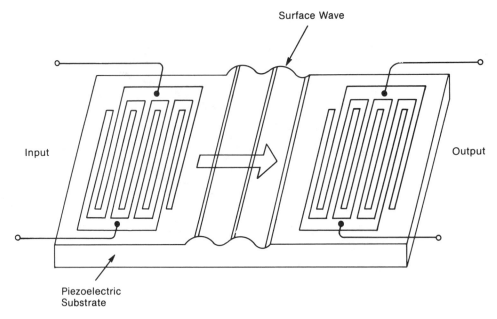

Figure 3.45 SAW filters are made on a piezoelectric substrate with sending and receiving transducers consisting of sets of interdigitated metal electrodes.

SAW filters are made on substrates of piezoelectric materials such as quartz or lithium niobate. To this is added a pattern of interdigitated metal electrodes which, when an AC voltage is applied between them, cause the substrate to oscillate. The spacing between electrodes determines the wavelength and, therefore, the frequency sent. The length of the electrode "fingers" determines the strength of the signal and the number of electrodes is inversely proportional to bandwidth. The final characteristics are very repeatable since the same photolithography process for integrated circuit fabrication is used to create the electrode patterns. A view of the substrate, sending transducer, and receiving transducer is shown in Figure 3.45.

SAW filters are excellent for color television receiver IF filters where phase and amplitude characteristics must be carefully controlled to provide a good picture. Radar receivers are also an excellent application for the filters since the characteristics can be closely "matched" to the pulses returning from the target.

3.10 REVIEW QUESTIONS

1. What kind of filter would you use if you wished to get rid of all signals below a certain frequency?

2. What kind of filter would you use to get rid of a narrow band of frequencies?

3. If a low-pass filter contains four reactive components, what is the final slope that the response settles down to in the stopband?

4. What happens if a filter designed for 100 ohm use is used with a 75 ohm source and load instead?

5. For the situations described in question 4, which would likely result in the poorer delay response and why?

6. What is the attenuation (dB), phase shift (degrees), and delay (seconds) at 6.2 MHz of a simple low-pass filter formed by placing a 740 pF capacitor across the connections from a 600 ohm source to a 700 ohm load (Figure 3.4 (a))?

7. What is the resonant frequency and 3 dB bandwidth of a parallel tuned circuit operating with a 2200 ohm source, 6800 ohm load, $L = 22 \mu H$, and $C = 470$ pF?

8. What is the attenuation (dB) and phase shift (degrees) at 1.2 MHz for the tuned circuit of question 7?

9. For the two-element, low-pass filter of Figure 3.12, what value of load resistor will result in maximum power transfer from source to load?

10. Design a three-element, low-pass filter with a flat amplitude response. It will have a total Q of 1.0 and a "resonant" frequency of 3.65 MHz. (The 3dB cutoff frequency will be about the same as the resonant frequency for this particular value of Q.) The source and load resistances will be 75 ohms.

11. Find the parallel equivalent of these impedances:

    a. $25 + j85$ ohms

    b. $130 - j50$ ohms.

12. Find the series equivalent of these impedances:

    a. 1000 ohms in parallel with $+j560$ ohms

    b. 860 ohms in parallel with $-j900$ ohms.

13. Design an impedance-matching circuit similar to that of Figure 3.24(a). The source is 75 ohms, the load is 1500 ohms, the operating

frequency is 12.0 MHz, and the bandwidth is to be 0.95 MHz. All components are assumed to be lossless.

14. You have a matching network similar to that of Figure 3.24(c) and with component values $C_1 = 500$ pF, $C_2 = 200$ pF, $L = 0.005$ μH. The load across the inductor is 500 ohms. What resistive input impedance would you measure across C_1?

15. Design a critically coupled bandpass filter similar to that of Figure 3.33. The center frequency is to be 1.5 MHz and the bandwidth is to be 50 kHz. The source and load resistances will each be 1200 ohms.

4. OSCILLATORS

An oscillator is a signal source that is used in radio transmitters, receivers, and data transmission circuits. Oscillators may produce any number of different waveforms but the most important for communication circuits is the sine wave. The circuits in this chapter are intended for operation at higher frequencies—above the audio range. There is no theoretical reason why they aren't useable at lower frequencies. It is simply that most use inductors and, at lower frequencies, these become rather large and, therefore, impractical.

This chapter will discuss:

1. Oscillator characteristics—what makes the difference between a good circuit and a bad one.

2. Oscillator fundamentals—what gets them started and keeps them running.

3. Three basic forms—an introduction to specific designs such as transformer-coupled, Colpitts, Clapp, and Hartley oscillator circuits.

4. Oscillator drift—why the operating frequency wanders and how to minimize the problem.

5. Practical oscillators—real designs that work.

6. Crystal control—for greater stability and accuracy.

7. Voltage-controlled oscillators—using varactor diodes.

4.1 OSCILLATOR CHARACTERISTICS

Before looking at oscillator circuits in detail we should examine the important characteristics of the circuits in general and the requirements of the circuits in which they will be used.

The most obvious characteristic is the frequency of oscillation. Most of the designs we will be looking at are sine wave generators that can be used from 10 kHz up to 1000 MHz, if the right components and transistors are used. Above that frequency, the same theory of operation is still valid but the circuits change drastically in physical appearance.

A second characteristic describes how carefully the oscillator "stays" at the desired frequency. There are two parts to this. The first problem is that oscillators tend to slowly drift or wander in frequency as either their operating temperature or supply voltage changes. Many will also change frequency as they get old. The second problem is a more rapid frequency jitter that could be caused by poor design or poor component selection. Figure 4.1 shows a sine wave that could have come from an oscillator. A wave such as this, viewed on an oscilloscope, won't show any problems unless they are very severe. The slow frequency drift could be detected with a good frequency counter. Notice that we are saying a "good" frequency counter because this instrument also contains an oscillator and it could drift too. The more rapid jitter can be seen only on a spectrum analyzer. A single frequency of oscillation should produce a single, thin, vertical line on the display whereas a jittering frequency will produce a wider line with what are called *sidebands*.

The drift problem means that if you jump into your car and turn on the FM radio (98 to 108 MHz) you would likely set the dial to one of your favorite radio stations. Then, as the car warms up and you drive over a few bumps, you may need to continually adjust the dial to stay on that station. Drift is more serious when the operating frequency is higher because the same percentage change in frequency means a greater number of kilohertz change than at the lower frequency. For this reason, special automatic frequency control (AFC) circuits are added to TV and FM tuners which both operate at frequencies

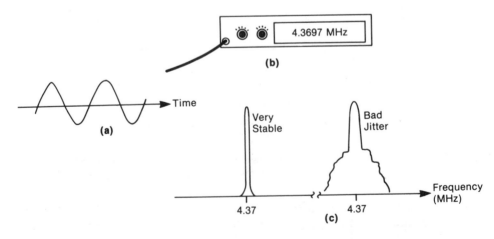

Figure 4.1   Sine wave as seen on an oscilloscope shows very few problems (a); a frequency counter (b) will indicate slow frequency drift, and a spectrum analyzer (c) will show the more rapid jitter.

above 50 MHz. AM radios (540 to 1600 kHz), on the other hand, operate at much lower frequencies and rarely experience a problem. For very critical circuits such as transmitters, quartz crystals in temperature-controlled packages are needed to control the drift. (Look back at Figure 3.38.)

A third requirement of an oscillator is its power output. Most designs are intentionally limited to less than about 20 milliwatts. This would be the rough equivalent of a 12 volt peak-to-peak sine wave across a 1000 Ω load resistor. Higher powers are avoided because they would heat the components and lead to drifting. Amplifiers placed after the oscillator can easily boost the power up to any desired level.

A final characteristic of an oscillator is its ability to be tuned to different frequencies. For some applications, an oscillator will spend its entire life at a single frequency. For other applications, the oscillator must be manually tunable to different frequencies. One of the circuits you adjust with the tuning control on your radio or TV receiver is an oscillator. A variation of the tunable oscillator is one that is tuned with a changing voltage rather than the mechanical action of turning the control of a variable capacitor or inductor. These voltage-controlled oscillators (VCOs) are used in data modems and frequency synthesizers.

4.2   OSCILLATOR FUNDAMENTALS

The important parts of any oscillator circuit are shown in Figure 4.2. The transistor acts as an amplifier to turn a little input signal into a larger output signal. Part of this output goes to the load as useful output energy from the circuit. The rest of the transistor's output goes into the network that consists of inductors and capacitors. The particular shape of the network shown in this illustration is only one of several possibilities, as we will see shortly. The network feeds its signal back to the input of the transistor and thereby provides the small signal that gets amplified to keep the oscillations going.

The network is the most critical part of the circuit as it determines the frequency of oscillation, how immune the circuit will be to temperature and other changes, and how the frequency can be tuned, whether it be manual or electrical.

An oscillator uses positive feedback to start the oscillators and keep them going. But, not all circuits that use positive feedback will oscillate. Active filters using op-amps are one example. They use positive feedback but don't oscillate, at least not if they work properly. To obtain oscillations, the amount of signal fed back to the input of the transistor must meet certain minimum conditions. The feedback network helps to determine this.

The minimum conditions necessary for oscillations are called the *Barkhausen Criteria*. This states that the signal fed back to the input must have the

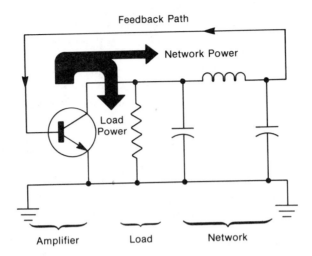

Figure 4.2 An oscillator has three parts: an amplifier, a load, and a feedback network.

same phase as the original input signal (positive feedback) and that it must have at least the same amplitude. This means that the gain of the transistor times the "loss" of the network must be at least 1.0. To use an example, suppose we had 1 milliwatt of signal going into the base of the transistor in Figure 4.2. If the power gain of the transistor is 10 then the output at the collector would be 10 milliwatts. Perhaps 9 milliwatts of this would go to the load as useful power output from the circuit and the remaining 1 milliwatt enters the network. If the network is made with only inductors and capacitors, it can't waste the power but only passes it on—assuming ideal components. The 1 milliwatt of signal should, therefore, come out the other end of the network and end up at the transistor's base.

The network "loss" that was just referred to isn't really a loss of energy inside the network; instead it represents the division of power between the load and the network at the collector side. Since one-tenth of the available power entered the network and the power gain of the transistor is 10 then the product of the two comes out to the magical 1.0. Figure 4.3 shows how this power division takes place. The two resistances are in parallel and so have the same collector AC voltage across them. The amount of current that flows into each will be inversely proportional to the resistance. Even though the network is made up of inductors and capacitors, its input impedance will be resistive since the other end is loaded down with the input resistance of the transistor.

Therefore, the power ratio will be:

$$\frac{P_{load}}{P_{network}} = \frac{R_{network}}{R_{load}} \qquad (4.1)$$

Remember that this magic number of 1.0 is an absolute minimum. In practice, a slightly larger signal is fed back to ensure that the circuit always works and to make up for small losses that do exist in any practical network.

To get the signal that comes out of the transistor back into phase with the original input signal to the base, the network must provide a phase shift of 180°. The transistor itself provides the additional 180° when the signal goes into the base and out the collector. This makes a total of 360° for the transistor plus network and results in positive feedback. For stable oscillations, this exact phase shift should only exist at one frequency; otherwise, the circuit would try oscillating at several frequencies simultaneously and the resulting output would not be a pure sine wave.

But how does an oscillator get started in the first place? Where does the initial signal that gets amplified come from? The source is inside the transistor used and all transistors, FETs, and vacuum tubes come equipped at no extra charge. This source is the white noise produced by the small resistances and random current paths in the semiconductor material. When amplifiers are designed, this added noise is a nuisance as we saw in the first chapter. But, for oscillators, it is very useful. Being "white" noise means that it contains all frequencies with equal amplitude and so it conveniently starts oscillators of any frequency. Although the noise amplitude is extremely small, the oscillations will build up gradually provided that the network feeds back a little more than the minimum required. This starting buildup is shown in Figure 4.4.

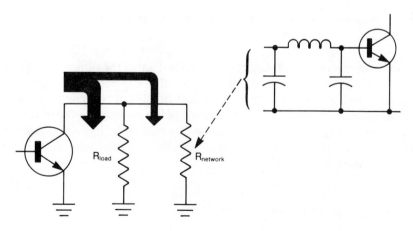

Figure 4.3   Power division at the collector of the oscillator's transistor.

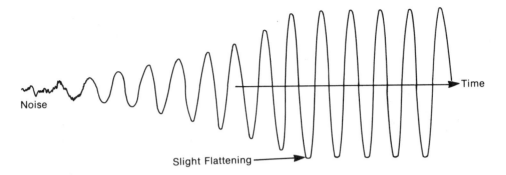

Noise

Slight Flattening

Time

Figure 4.4 Oscillator start-up begins with small amounts of random noise and ends when some slight distortion reduces the average gain.

The oscillations keep building up in amplitude until something stops them. That something is usually the transistor going into saturation or cutoff or both—depending on the biasing. During either of these flattenings, the gain of the amplifier will be temporarily reduced. This reduction, when averaged over the full cycle of the waveform, will be just enough to reduce the average gain down to 1.0. The oscillations will then stay at this amplitude. The extent of the flattening depends on how much larger than 1.0 the initial starting gain was. The designer is then faced with a problem; a high value of starting gain will guarantee that the circuit will always get going and that oscillations will build up rapidly. However, the output wave will be slightly distorted. On the other hand, a starting gain much closer to 1.0 might not start some morning (until it has its third cup of coffee).

4.3 THREE BASIC FORMS

Actual oscillator circuits can take many different forms, the differences appearing mainly in the feedback network. Remember that the network must produce a 180° phase shift at only one frequency and be able to transfer a small but predictable amount of power back to the base of a transistor. Three of the more common forms of high-frequency oscillators are described in the following sections. Only the essential parts are shown; complete circuits with all bias resistors, etc., will be shown later.

4.3.1 Transformer-Coupled Oscillator

The oscillator circuit of Figure 4.5 uses a transformer to feed back enough energy at the right phase to start and maintain oscillations. The resonant circuit in

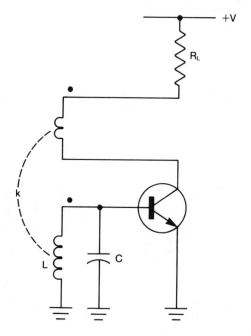

Figure 4.5 Transformer-coupled oscillator with a tuned secondary circuit.

the secondary circuit determines the frequency of oscillations. The formula for this is given below. The inductance in this case is the secondary inductance of the transformer winding.

$$F_{osc} = \frac{1}{2\pi\sqrt{LC}} \tag{4.2}$$

The 180° phase shift required by the network is obtained by selecting the right lead polarity on the transformer. The dots on the schematic are used to indicate this. If, for example, the collector current of the transistor increases, the collector side of the primary would become more negative than the dot. (The dot would be more positive than the collector.) When the signal reaches the tuned secondary, the dot there would also become more positive. This would make the base of the transistor more positive and so increase the base current and thereby increase the collector current. We are now back to where we started (more collector current) and so the connection is providing positive feedback. This condition will only exist at the resonant frequency of the secondary tuned circuit. Above and below resonance, the secondary signal would be both weaker and also shifted in phase one way or the other. Either condition is enough to stop oscillations. The amount of signal fed back is controlled by the turns ratio of the transformer and also by the coupling coefficient (k—that

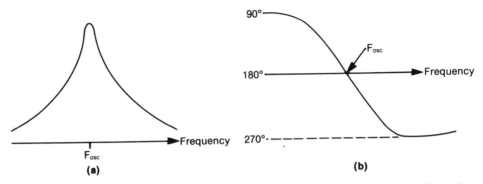

Figure 4.6 Amplitude (a) and phase response (b) of the tuned transformer used to set the operating frequency of the transformer-coupled oscillator.

was discussed back in Chapter 3). The amplitude and phase response of the tuned transformer are shown in Figure 4.6.

Some variations on this circuit are possible. To make the oscillator variable, the capacitor can be paralleled with a variable one or the transformer could be equipped with a moveable core. This happens to be the method used to tune the oscillator for automotive AM receivers to different stations. The transformer is part of the assembly shown back in Figure 2.14.

Another variation of the circuit is to move the tuned circuit from the base side of the transformer up to the collector side. As long as the right relation between amplitude and phase shift is retained, the actual location of the tuned circuit makes little difference.

4.3.2 Colpitts Oscillator

A second type of oscillator is the Colpitts circuit shown in Figure 4.7. This is a very common circuit and is the one we used as the introduction to this chapter. The network is a three-element, low-pass filter similar to those discussed in Chapter 3. The amount of signal that is fed back to the base is controlled by the ratio of the two capacitors. They determine how much larger the input impedance to the network at the collector side is than the transistor's input resistance at the base. Remember that it is the relation between the network's resistance and the load resistance that determines how much signal is fed back. We will look at an example of this shortly.

The components of the network are selected to give a moderately high loaded Q (about 10) and so the frequency response of the network ends up with a pronounced peak as shown in Figure 4.8. The peak also happens to be the point at which the network creates a phase shift of 180° so this is the frequency at which the oscillations occur.

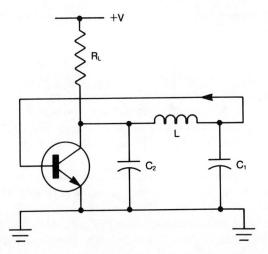

Figure 4.7 Colpitts oscillator uses a low-pass filter for a feedback network.

The peak frequency and, therefore, the oscillating frequency is given by:

$$F_{osc} \approx \frac{1}{2\pi \sqrt{LC_{equiv.}}} \tag{4.3}$$

$$where \quad C_{equiv} = \frac{C_1 \times C_2}{C_1 + C_2} \tag{4.4}$$

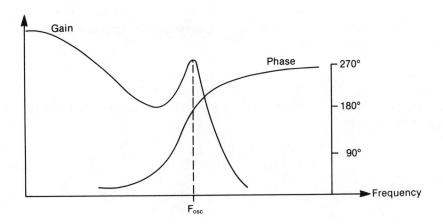

Figure 4.8 Amplitude and phase response of the Colpitts network.

This is an approximation because it depends somewhat on the exact value of loaded Q used with the network.

The three-element, low-pass filter will have a phase shift that starts at $0°$ at low frequencies and increases up to a maximum of $270°$ at very high frequencies. That is a maximum phase shift of $90°$ for each reactive component. There will, therefore, be a frequency between these extremes where the phase shift will be exactly $180°$ and this is the oscillating frequency. If the network Q is kept high then the phase shift curve passes through the $180°$ point at a steep slope and this tends to keep the oscillator operating at one frequency rather than wandering.

This circuit can also be made tunable. The inductor or the capacitor can be varied. Since C_2 is usually the smaller value, it is usually the one that is varied rather than C_1.

Keep in mind that the circuit so far is not complete. The transistor has no bias resistors at its base. In fact, there is a DC connection directly from the collector back to the base through the inductor. Practical designs will be looked at later.

Now let us look at an example. This will help to describe the relation that must exist between the components that make up the network, the load resistance, and the input resistance of the transistor.

Example 4.1

Use the basic Colpitts diagram of Figure 4.7 and the following values:

$$L = 5.5\mu H, \ C_1 = 6800pF, \ C_2 = 470pF, \ R_{load} = 1000\Omega$$

The input resistance of the transistor is 200 ohms. What is the frequency of oscillation, and what minimum power gain does the transistor need to make sure that the oscillations get started?

Solution:

First we will find the oscillating frequency. The equivalent value of the two capacitors in series is:

$$C_{equiv} = \frac{C_1 \times C_2}{C_1 + C_2}$$

$$= \frac{6800 \times 470}{6800 + 470}$$

$$= 439.6 \text{ pF}$$

The resonant frequency of the tuned circuit is:

$$F = \frac{1}{2\pi \sqrt{LC_{equiv}}}$$

$$= \frac{1}{2\pi \sqrt{5.5 \times 10^{-6} \times 439.6 \times 10^{-12}}}$$

$$= 3.237 \text{ MHz}$$

This should be very close to the oscillating frequency.

Now, to find the power gain that is needed, we must find how the transistor's output power splits between the network and the load. Our first job is to find the input impedance of the network.

$$Z_{network} = \left(\frac{C_1}{C_2}\right)^2 \times R_{in \text{ transistor}}$$

$$= \left(\frac{6800}{470}\right)^2 \times 200$$

$$= 41.87 \text{ k}\Omega$$

(This formula is a variation of the ones used for impedance matching back in Figure 3.24.)

The output power from the transistor will divide as follows:

$$\frac{P_{load}}{P_{network}} = \frac{Z_{network}}{R_{load}} = \frac{41.87k}{1 \text{ k}} = 41.87$$

For each milliwatt of power going into the network (and therefore back to the input of the transistor) there will be another 41.87 milliwatts going into the load. The total output power from the transistor would have to be:

$$P_{out} = P_{network} + P_{load}$$

$$= 1 \text{ mw} + 41.87 \text{ mW}$$

$$= 42.87 \text{ mW}$$

Therefore, the minimum power gain of the transistor would be:

$$\frac{P_{out}}{P_{in}} = \frac{41.87}{1} = 42.87$$

If you happen to want the gain in decibels:

$$\text{Power gain (dB)} = 10 \log \left(\frac{P_{out}}{P_{in}} \right)$$

$$= 10 \log 42.87$$

$$= 16.3 \text{ dB}$$

A variation of the Colpitts circuit is called the *Series Colpitts* or the *Clapp oscillator*. The inductance of the previous circuit now becomes a combination of a series inductor and capacitor. The combination must be inductive at the oscillating frequency and so this will be above the series resonant frequency of the two components.

The advantage of this circuit over the regular Colpitts circuit is improved frequency stability; the oscillating frequency shouldn't wander around as much. This improvement is caused by the much steeper phase shift versus frequency curve that results from operating just above the series resonant frequency of L and C_3.

The circuit in Figure 4.9 includes some component reactances as a further example of the operation of an oscillator. Notice how the reactances of C_1 and C_2 add up to $-j112\Omega$ and the reactances of L and C_3 do the same but with the opposite sign ($+ j112\Omega$). This makes the network a resonant circuit and, therefore, these should be the conditions at the operating frequency of the oscillator. With the 1000Ω load resistor used, the loaded Q of the network should be:

$$Q_{loaded} = \frac{R_{load}}{X_{C_2}} = \frac{1000}{-j100} = 10$$

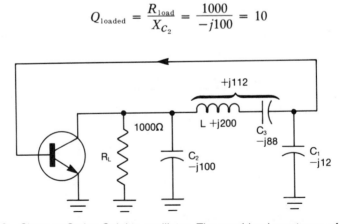

Figure 4.9 Clapp or Series Colpitts oscillator. The combined reactance of L and C_3 must be inductive at the oscillating frequency.

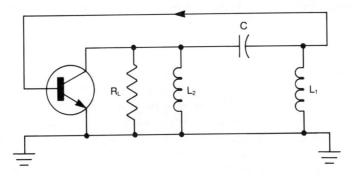

Figure 4.10 Hartley oscillator uses a high-pass filter for a feedback network.

4.3.3 Hartley Oscillator

The third oscillator circuit in popular use is the Hartley oscillator with the basic circuit of Figure 4.10. The network used for this oscillator circuit is also a three-element filter but, in this case, it is a high-pass filter. At very low frequencies, the phase shift will reach a maximum of 270° and at very high frequencies the shift will decrease to zero. Therefore, as with the low-pass filter, there will be a frequency between these two extremes where the phase shift will be exactly 180°. This will be the oscillating frequency—as long as the gain requirements are properly looked after. The amplitude and phase response of this network will be a mirror image of that shown in Figure 4.8, just reverse the direction of the frequency axis.

The amount of signal fed back to the transistor's input is controlled by the ratio of L_1 to L_2. As before, this ratio, along with the input resistance of the transistor, determines the network input impedance. The ratio of this impedance and the load resistance then set the minimum gain needed for the oscillator. Normally, L_1 is the smaller of the two inductors. The approximate oscillating frequency is given by:

$$F_{osc} \approx \frac{1}{2\pi\sqrt{(L_1 + L_2)\,C}}$$

The circuit doesn't really need two inductors. One inductor with a tap for the "ground" connection is normally used in any practical circuit. The inductance value in the formula would then just be the total inductance of the single coil.

4.4 OSCILLATOR DRIFT

We have mentioned that one of the biggest problems with good oscillator circuits is that their frequencies tend to drift. What causes this problem and how do designers control it?

Drift can occur for a number of reasons. Part of the problem is the transistor itself. In Chapter 2 we saw that any transistor contains a number of internal capacitances and these change their value when either the temperature or the bias voltages change. At frequencies above about 100 kHz, the capacitances cause additional phase shift inside the transistor. As a result, the signal at the collector of a transistor is usually shifted a bit more than the normal 180°. This means that the network need only add a bit less than 180° to maintain the overall 360° needed for positive feedback. As a result, the Colpitts circuit would operate at a slightly lower frequency than expected (look at Figure 4.8 again) and the Hartley would operate at a slightly higher frequency. The oscillating frequency formulas have always been given as approximate values because the extra shift inside the transistor is fairly difficult to calculate.

The second cause of frequency drift is the fault of the network itself. Most inductors and capacitors will change their value as temperature changes. In fact, some poor quality capacitors will also change value if the DC voltage across them changes. In the case of inductors, a temperature increase will cause the copper wire to expand in length and so increase the physical diameter of the coil slightly. The inductance, therefore, increases slightly. This relationship is called a *positive temperature coefficient* since the temperature increase causes an inductance increase in the same direction. Most capacitors also have a positive temperature coefficient that just aggravates the drift problem. However, special capacitors are available with predictable values of negative temperature coefficient (NTC). A clever circuit designer will use these capacitors to cancel out the inductor's drift. An important point to the student—if you are ever servicing an oscillator circuit, don't just replace capacitors with a similar value. Very often an "exact" replacement is needed.

In summary, it is very easy to build a circuit that will oscillate but it is very difficult to make a good, stable oscillator. Some of the techniques used are:

1. Carefully regulated and filtered power supplies.
2. Shielding of the oscillator circuit.
3. Temperature control of the oscillator compartment.
4. Temperature-compensated components used in the feedback network.
5. Mechanically rigid construction to prevent component movement from shock and vibration.

4.5 PRACTICAL OSCILLATORS

The circuits that we have been looking at so far were not practical circuits. They didn't have proper power supply connections or biasing components. This section examines three examples of working circuits. Be careful when you are looking at these circuits because they often look totally different from their basic circuit. The differences will be explained.

A full schematic for a 25 MHz Colpitts oscillator is shown in Figure 4.11. The two base resistors form a voltage divider that creates a base voltage of 2.5 volts (relative to ground). The emitter voltage will be about 0.7 volts lower than this, leaving 1.8 volts across the 180 Ω emitter resistor. The emitter current will be 10 milliamps as a result. Because collector and emitter current are about the same in a high-gain transistor, the collector current will also be 10 milliamps. (Remember that we are considering DC bias current and DC current gain rather than high-frequency AC current gain at the moment.) Since there will be no appreciable voltage drop across the collector inductor, the DC collector voltage will be the same as the power supply or 9.0 volts. The collector-to-emitter bias voltage is then 9.0–1.8 or 7.2 volts.

The AC part of the circuit can be viewed in two different ways. Some people see the transistor connected as a common emitter amplifier. This approach corresponds to the diagram of Figure 4.7 in our introduction to the basic Col-

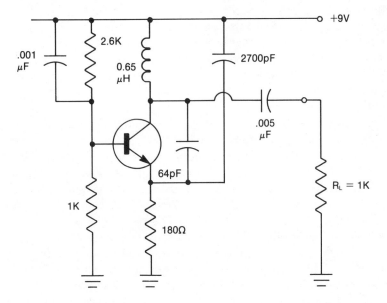

Figure 4.11 25 MHz Colpitts oscillator circuit with all bias components added.

pitts circuit. This is the approach favored by the author as students are generally more familiar with the common emitter form than any other. Using this approach, the circuit operation is as follows. At the collector of the transistor, a large part of the signal goes into the 1k Ω load resistor and a smaller part goes into the feedback network which will eventually get the signal back to the base. The network consists of the 0.65 μH inductor and the 65 pF and 2700 pF capacitors. The two capacitors, when considered in series, will resonate with the inductor at the 25 MHz operating frequency. The signal leaves the network at the top of the inductor, which just happens to be the positive power supply connection. From there the signal goes to the base through the low reactance of the 0.001 μF capacitor. This may seem like a very strange path to follow but it does make sense. Remember that we are considering the emitter of the transistor as the reference (common emitter amplifier) and the emitter isn't grounded. An ordinary common emitter amplifier usually has a capacitor from emitter to ground but this circuit doesn't. Therefore, it is quite alright to consider an AC voltage at the positive power supply relative to the emitter.

The other possible way of viewing the circuit is to consider the transistor as a common base amplifier. This is the approach that you will have to take if you can't accept the path the signal takes through the power supply to the base. The base, in this case, is considered as being at AC ground and the feedback path is then from the collector back to the emitter. The network would then be viewed in a different configuration and would send a noninverted signal back to the emitter.

Either view is correct, although the first is preferred. But, how is it possible to use two transistor configurations to look at the same circuit? After all, amplifier circuits are viewed either one way or the other but not both. The reason we have the choice for this and any other oscillator circuit is that oscillators don't have any input signal; they produce the signal themselves. An oscillator only has an output and so we are free to consider the "input" signal as following any path that is convenient.

Our second practical circuit is a Hartley oscillator. The full schematic for a 10 MHz tunable oscillator is shown in Figure 4.12.

The explanation of this circuit is very similar to the Colpitts circuit just described. The biasing is identical. Two capacitors are involved in the tuning circuit but they are connected in parallel and so act as a single component. The combination of the fixed and the variable value gives the oscillator a tuning range of about 15%, from 9.0 to 10.5 MHz. The resonance of these capacitors and the 1.5 μH inductor determines the operating frequency.

The feedback ratio is set by the position of the tap on the inductor. The closer the tap is to the positive power supply connection, the smaller the signal fed back to the base. Although it may not be obvious, the portion of the inductor from the tap up to the supply voltage is actually in parallel with the base and it is this portion of the inductor voltage that is used for the transistor input. The connection back to the base is through the power supply filter capacitor,

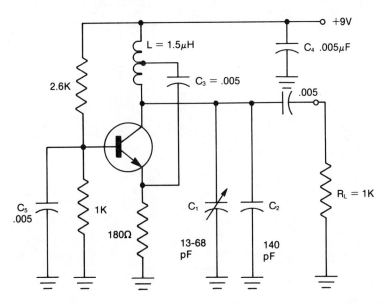

Figure 4.12 10 MHz tunable Hartley oscillator circuit with all bias components added.

C_4, and the base bypass capacitor, C_5. The base capacitor could also have been connected to the positive supply line as it was for the Colpitts circuit. Both the ground line and the positive supply line are considered as "AC ground."

The capacitor from the tap on the inductor, C_3, down to the emitter is also a low-reactance bypass capacitor that simply "grounds" the inductor tap to the circuit reference point—the emitter. This breaks the inductor into the two components shown in the basic circuit. It would be a good idea for the student to try redrawing this circuit to see if it can be matched to the basic Hartley circuit of Figure 4.10.

A third schematic of a working oscillator is shown in Figure 4.13. This is a series Colpitts or Clapp oscillator. Instead of a bipolar transistor, a field effect transistor is used just to show some variety. The component designations are the same as used on the basic diagram so that comparisons can be made. The FET, in this case, looks like it is connected in a common drain configuration (the FET equivalent of an emitter follower or common collector amplifier for those more familiar with bipolar circuits). However, remember that an oscillator has no input so we can select any configuration we want if it makes it easier to understand the circuit. To match the basic diagram, we need to find a connection from the drain of the FET back to the input of the network. This is provided through the power supply connection and the filter capacitors in the power supply. The source is considered as the reference point for this analysis. The ratio of C_1 and C_2 determines the amount of signal fed back to the input of the transistor for amplification and the series value of C_3 and the inductor operate with the other two capacitors to determine the oscillating frequency. At this

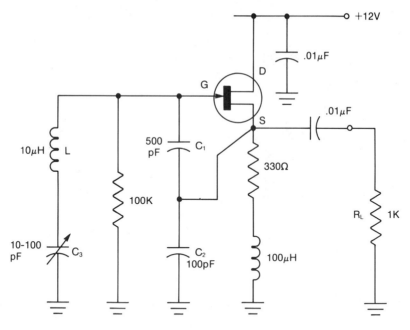

Figure 4.13   Series Colpitts or Clapp oscillator tunable to 10 MHz.

frequency, C_3 and the inductor must appear as an equivalent inductive reactance. The placement of C_3 is very handy since one side of variable capacitors is usually grounded.

The 100kΩ resistor at the gate of the FET and the 330Ω resistor at the source form the bias circuit for the transistor. The extra 100μH coil in series with the resistor isn't vital but is often used to increase the total impedance of the bias circuit without upsetting the DC values. This is beneficial because, without the coil, the 300Ω resistor would be directly in parallel with the output connection to the load and so would waste some of the oscillator's output power. It would also be in parallel with part of the feedback network, C_2, and so would lower the Q of the network with the result that the oscillator might be less stable and more prone to wandering around in frequency. This particular technique can be used on any of the other circuits. One danger is that it might tend to form an oscillating frequency of its own, but, careful design would avoid that.

4.6   CRYSTAL CONTROLLED OSCILLATORS

If a mobile radio transceiver (transmitter and receiver in one package) is assigned a frequency of, for example, 155.000 MHz, it must remain very close to that frequency even though the truck it is mounted in operates in the cold of winter and the heat of summer and over rough roads. The problem is that other mobile units may be assigned operating frequencies 15 kHz above and below the truck's frequency and these must not be interfered with. If we arbitrarily allow the truck's frequency to be in error by 5 kHz, we can get an idea of the stability needed by the oscillator.

$$\% \text{ stability} = \frac{5 \text{ kHz}}{155.000 \text{ MHz}} \times 100$$

$$= .0032\%$$

This could also be referred to as a stability of 32 parts per million or, more simply, 32 ppm.

This kind of stability is far beyond the capability of any of the oscillator circuits we have been looking at so far. They can all be easily adjusted to the right frequency and they might even remain there for 10 minutes or so but then they would slowly wander off to some other frequency.

The solution to the accuracy and stability problems (these are two separate problems) is the quartz crystal. This is the same device that was introduced in Chapter 3 to make very narrow bandpass filters. The equivalent circuit and its impedance variation over a wide range of frequencies is shown again in Figure 4.14 to refresh your memory. Remember that the quartz crystal has two valuable properties. First, its Q is extremely high ($> 10,000$) and second, its characteristics are extremely stable.

Back in Figure 3.38 we saw that if the slices were cut out of the original crystal at a very precise angle then the resonant frequency will also stay very

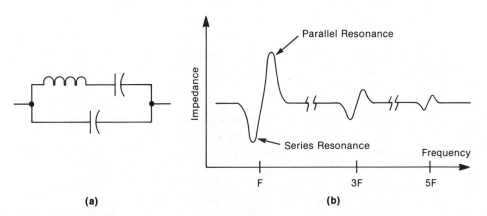

Figure 4.14 Equivalent circuit of a quartz crystal (a) and its impedance variation over a wide range of frequency (b). Notice that a typical crystal has weaker resonances at frequencies that are odd multiples of the main frequency.

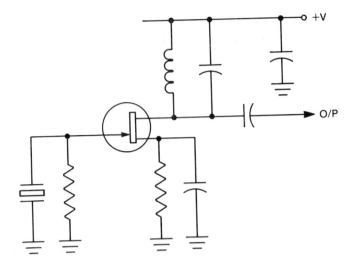

Figure 4.15 Crystal-controlled Miller oscillator.

constant over a wide range of temperatures. Keeping in mind that the crystal is simply a high-quality tuned circuit with a dominant series and parallel resonant frequency that are very close together, let us see how it can be added to an oscillator circuit to improve its stability. One of the simplest circuits is shown in Figure 4.15.

This circuit is often called a *Miller oscillator circuit*. It could also be designed with a bipolar transistor or even a triode vacuum tube. The quartz crystal is connected from gate to ground and the tuned circuit is connected from the transistor's drain to the positive supply voltage which can also be considered as an AC ground. At first glance there doesn't appear to be any feedback path from output back to the gate. How does the circuit oscillate?

The answer is found inside the transistor. As we have already explained, all transistors have internal junction capacitances. For the FET, in this case, the drain-to-gate capacitance is the one that provides the feedback path. This is another example of an internal problem of a transistor that is occasionally useful. The capacitance is usually less than about 1 pF however, and for low-frequency oscillators, it may not be large enough to allow enough signal to feed back. In this case, an external capacitance can be added.

The equivalent circuit of the Miller oscillator at its operating frequency is shown in Figure 4.16. This illustrates how the circuit components form a three-element, high-pass filter that can shift the output signal phase by the necessary 180° for the transistor's input. The tuned circuit at the drain of the transistor will act as an equivalent inductor and so must be tuned for resonance slightly above the desired oscillating frequency. The crystal is also shown as an equivalent inductance and so it must be operating some place between its series and

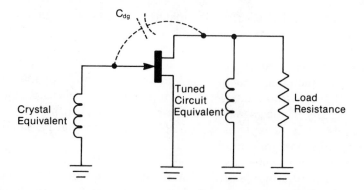

Figure 4.16 Equivalent circuit of the Miller crystal oscillator. Both the tuned circuit and the quartz crystal will be inductive at the oscillating frequency.

parallel resonant frequencies. The distance between these two resonances is, of course, very small so the frequency of oscillation can be predicted fairly accurately; simple tuning adjustments can usually make the frequency exactly what was desired. The equivalent circuit shows that this oscillator is actually very similar to the Hartley circuit.

A second crystal-controlled oscillator is shown in Figure 4.17. This particular circuit was chosen to illustrate several important features and problems.

This oscillator is based on the Colpitts circuit of Figure 4.13 and is designed with a bipolar transistor. The only tuned circuit consists of the crystal

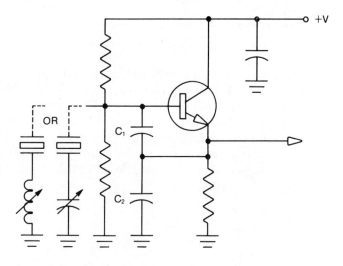

Figure 4.17 Tunable crystal oscillator based on the Colpitts circuit.

reactances and the two extra network capacitors, C_1 and C_2. (Ignore the two variable components for the moment.) While the lack of any extra tuned circuit and, in particular, an extra inductor, may seem an advantage, it does mean that the designer must be very careful. Back a few pages, Figure 4.14 showed that all crystals have more than one resonance point and these tend to be more or less harmonically related. Without any extra L-C tuned circuits, this oscillator could try to operate at any of these "harmonics" or "overtones." Adjustment of the C_1 /C_2 ratio and choice of crystals with a strong fundamental resonance will improve the operation of this circuit.

Some crystals are specified as *overtone* crystals and these are intended for operation at the third harmonic of the fundamental resonance for some crystals and at the fifth harmonic for others. Fundamental crystals usually operate up to about 20 MHz and that is the upper limit of this circuit. Overtone crystals should not be used without an L-C tuned circuit to ensure operation at only one frequency.

The other feature of this circuit is the use of a variable capacitor or inductor to adjust the frequency of oscillation slightly. This circuit, like the previous one, operates between the series and parallel resonance points of the crystal. This is the inductive region. The added series components will either add to the equivalent inductance in the case of the variable inductor, or subtract in the case of the variable capacitor. The frequency can, therefore, be varied by a small amount. However, remember that the equivalent inductance of the crystal is huge and that the oscillating frequency must stay between the two resonance points. Therefore, the range of tuning is very small—less than 0.2% in most cases.

4.7 VOLTAGE-CONTROLLED OSCILLATORS

The previous section was concerned with oscillators that had to operate at one precise frequency. Now we look at circuits that can be electrically (rather than manually) tuned over a wide range of frequencies. The result is called a *voltage-controlled oscillator* (VCO). Such a circuit is useful in a number of places. One application is in a frequency-modulated (FM) transmitter where the music or voice signal must cause the transmitted carrier frequency to vary in step with the signal amplitude. A second application is found in the tuner of any television receiver with pushbutton tuning.

4.7.1 Varactor Diode

The most common method of voltage tuning uses the varactor diode. This is a diode operated with reverse bias so that it doesn't conduct but rather acts as a variable capacitor. The word *varactor* is made up from its description "voltage variable capacitor." A varactor diode is not very different from any other diode; it is just made with more emphasis on the capacitance properties in the reverse direction than on the rectifying properties in the forward direction. All semiconductor junctions have this variable capacitance property including the junction capacitances inside all transistors. The diodes sold as varactors will have a more carefully controlled capacitance curve so that they can be used in production line assembly work and lower losses so that the capacitance will have higher Q.

The variable capacitance results when a reverse bias voltage pulls the charge carriers away from the junction leaving an area depleted of electrons and holes. This is shown in Figure 4.18(a). The depleted area is, therefore, an insulator. The region within the charges acts the same as the charged metal plates of a capacitor. The result is two charged regions separated by an insulator—the same condition that forms a charged capacitor. When the reverse bias voltage increases, the charges move further apart. This is the same as moving the plates of a capacitor apart so the junction capacitance goes down. Lowering the reverse bias will increase the capacitance. The graph of Figure 4.19 shows this variation of capacitance with reverse voltage for an assortment of diodes. Diodes are available with a wide range of capacitances and a difference in the shape of the capacitance curves is also available. The more rounded curves in the figure are the result of manufacturing diodes with a very abrupt change from the N region of the semiconductor material to the P region. This results in a greater capacitance change for a given voltage change. Note carefully that both the vertical and the horizontal axes of the graph have logarithmic scales and this makes the curves look more linear than they actually are.

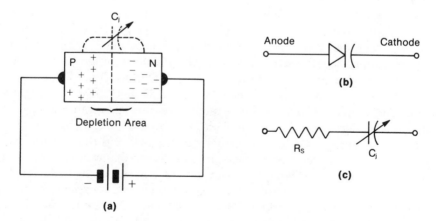

Figure 4.18 Location of charges inside a reverse biased diode (a), varactor diode symbol (b), and its equivalent circuit (c).

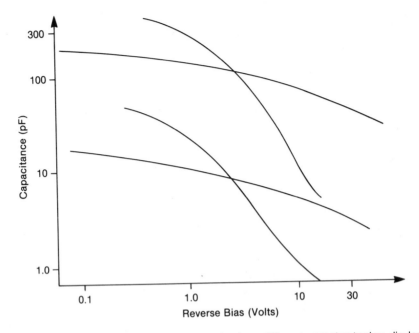

Figure 4.19 Capacitance vs. voltage curves for four different varactor tuning diodes.

Variable capacitance diodes are cataloged on data sheets according to their capacitance at −4 volts. One manufacturer, for example, has values available from about 5 pF up to 400 pF. This is just the value at −4 volts; by changing the bias, each diode will usually increase and decrease its capacitance by a factor of two or three in each direction. Data sheets will also include a value for the Q of each diode. Typical values at −4 volts range from 100 up to 500. However, the Q also changes with reverse-bias voltage, temperature, and frequency so that a designer will have his or her hands full if he or she is trying to make an exact calculation. The inductor that is used with the tuning diode is usually lower in Q though and so determines the overall Q in most applications.

Now, how does the tuning diode or varactor diode change a resonant frequency? The simplest application to a tuned circuit is shown in Figure 4.20. A reverse-bias voltage is applied to the diode through a large value resistor. Since the diode is reverse bias, it draws virtually no current and the resistor causes very little change in the bias voltage. It is necessary, however, because the power supply would otherwise place a low impedance across the capacitance of the diode and so lower the Q of the whole tuned circuit. A capacitor is added between the diode and the inductor to block DC current from flowing through the inductor. Without the blocking capacitor, it would be impossible to develop any reverse bias across the diode. The result is a tuned circuit whose resonant frequency is variable with changes in the DC bias voltage. The resulting curve

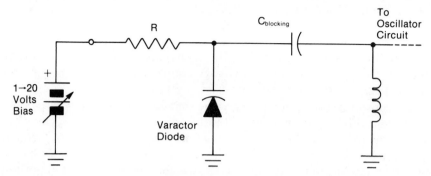

Figure 4.20 Variable frequency resonant circuit. The series resistor prevents the power supply from lowering the circuit Q. The blocking capacitor prevents the inductor from shorting out the bias voltage.

of bias voltage versus resonant frequency is shown in Figure 4.21. These curves were calculated assuming a 10 μH inductor and the two larger-value diode characteristics from Figure 4.19. Notice that while the horizontal voltage scale is still logarithmic, the vertical frequency scale is now linear. The curves are now a little less "nonlinear" since the resonant frequency is inversely proportional to the square root of the capacitance. In fact, the diode with the very abrupt junction comes close to having a linear relation of voltage versus frequency. (It still looks curved in Figure 4.21 because of the logarithmic scale.)

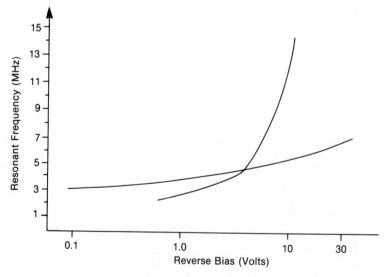

Figure 4.21 Frequency vs. tuning voltage for a varactor-controlled resonant circuit. The very abrupt junction diode has the wider tuning range.

4.7.2 Varactor-Controlled Oscillator

The varactor tuning diode can be used to vary the frequency of any of the oscillators discussed so far—including the crystal oscillators (over a very narrow frequency range). An example of a voltage-controlled Colpitts circuit is shown in Figure 4.22.

The control voltage is a positive value that can change anywhere between 1 and 20 volts. The higher voltage will result in a smaller junction capacitance for the diode and this will make the oscillator operate at its highest frequency. The relation between frequency and voltage is not linear; although, with careful design, the relation can be very close to a straight line over a limited portion of the oscillator's range.

The components that determine the operating frequency are shown in Figure 4.23. The four capacitances form a series-parallel combination that can be reduced down to one equivalent value as shown in (b). A more detailed analysis would include the input capacitance of the transistor. This wouldn't be an easy task because the total input capacitance, as mentioned before, depends on the gain of the transistor at the oscillating frequency. The calculation is only important at higher frequencies where the value of C_1 is smaller and, therefore, the transistor's capacitance would be more significant.

One important characteristic of any VCO is the highest and lowest frequency that it will operate at. This can be found by using the equivalent circuit and making two calculations—one for the minimum diode capacitance and the other for the maximum capacitance. The following example illustrates this.

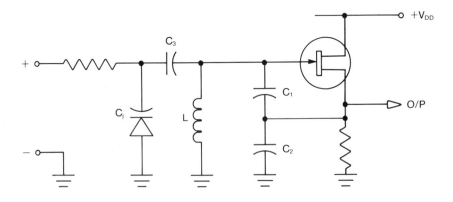

Figure 4.22   Voltage-controlled Colpitts oscillator.

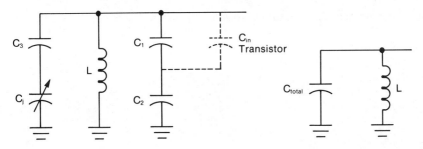

Figure 4.23 These components determine the frequency of oscillation for the Colpitts voltage-controlled oscillator. Their equivalent circuit is shown in (b).

Example 4.2

A voltage-controlled oscillator uses the circuit shown in Figure 4.22. Its component values are:

$$L = 10 \ \mu H$$
$$C_1 = 100 \ pF$$
$$C_2 = 50 \ pF$$
$$C_3 = 270 \ pF$$

Without considering any extra capacitance for the transistor, calculate the highest and lowest oscillating frequency if the varactor changes its junction capacitance from a minimum of 4 pF up to a maximum of 40 pF.

Solution:

Using Figure 4.23 as a starting point, we can reduce this down to an equivalent circuit with one capacitor and one inductor. This will have to be done twice— once for the 4 pF and once for the 40 pF value of varactor capacitance.

C_1 and C_2 are in series. Their combined value is:

$$= \frac{C_1 \times C_2}{C_1 + C_2}$$

$$= \frac{100 \times 50}{100 + 50}$$

$$= 33.33 \ pF$$

C_3 and the junction capacitance are also in series. Their combined capacitance is:

For minimum varactor capacitance:

$$= \frac{C_3 \times C_j}{C_3 + C_j}$$

$$= \frac{270 \times 4}{270 + 4}$$

$$= 3.94 \text{ pF}$$

For maximum varactor capacitance:

$$= \frac{C_3 \times C_j}{C_3 + C_j}$$

$$= \frac{270 \times 40}{270 + 40}$$

$$= 34.85 \text{ pF}$$

The two equivalent capacitances can now be placed in parallel to produce the final equivalents—one for the minimum capacitance condition and one for the maximum.

For minimum varactor capacitance:

$$C_{total} = 33.33 + 3.94$$
$$= 37.27 \text{ pF}$$

For maximum varactor capacitance:

$$C_{total} = 33.33 + 34.85$$
$$= 68.18 \text{ pF}$$

Now, finally, with the minimum and maximum values of the total capacitance and the single value of the inductor, we can calculate the highest and lowest resonant frequencies.

The highest frequency (using minimum capacitance):

$$F_{osc} = \frac{1}{2\pi \sqrt{LC}}$$

$$= \frac{1}{2\pi \sqrt{10 \times 10^{-6} \times 37.27 \times 10^{-12}}}$$

$$= 8.244 \text{ MHz}$$

The lowest frequency (using maximum capacitance):

$$F_{osc} = \frac{1}{2\pi\sqrt{10 \times 10^{-6} \times 68.18 \times 10^{-12}}}$$

$$= 6.095 \text{ MHz}$$

The oscillator is, therefore, tunable from 6.095 up to 8.244 MHz.

4.8 REVIEW QUESTIONS

1. What are the three main components of an oscillator? Which one is the most critical?

2. If an amplifier has a power gain of 18 and 5% of the output signal is fed back to the transistor's input using positive feedback, will the circuit oscillate?

3. What is the advantage of having a high Q feedback network instead of a low Q one?

4. Why do crystal oscillators usually need LC tuned circuits even though the crystal itself could be considered as a tuned circuit?

5. If a transistor to be used for an oscillator circuit has a power gain of only 18 (not dB) and the load resistance for the oscillator must be 75 ohms, what is the highest value of network input resistance that can be used and still have the circuit oscillate (see Figure 4.3)?

6. For an oscillator of the type shown in Figure 4.5, what is its operating frequency if $L = 0.5$ mH and $C = 0.047$ μF?

7. What might happen if you were constructing the oscillator of Figure 4.5 and (a) the inductance leads were accidentally interchanged? (b) the capacitor leads became interchanged?

8. For a Colpitts oscillator similar to that of Figure 4.7, the component values are $C_2 = 180$ pF, $C_1 = 1000$ pF, and $L = 6.8$ μH. What is the approximate oscillating frequency?

9. Why might the correct frequency be slightly different from what you have calculated in question 8?

10. Design the network for a Colpitts oscillator at 5.90 MHz so that C_1 is eight times the size of C_2. The network should have a loaded Q of 10

and the load resistance in this case is 560 ohms. (Network Q is mainly caused by the parallel combination of C_2 and the load resistance.)

11. For a varactor diode that has a range of 8 pF to 55 pF, what is the ratio of the highest to lowest frequency that can be tuned?

12. For the same varactor diode as in question 11, what will the highest and lowest resonant frequency be for a parallel combination of the varactor, a 10 pF fixed capacitor, and a 10 μH inductor?

5. AMPLITUDE MODULATION

Modulation involves making changes in some way to a high-frequency carrier so that it represents, for example, music, voice, or data signals. If done properly, the resulting high-frequency signals can be transmitted over long distances and then demodulated back to create music, voice, or data signals identical to the original.

Why is modulation used? This was partially discussed in Chapter 1. Modulation makes it possible to send information over cable or antenna systems with much greater ease and with less noise distortion and data errors. It also makes it possible to select different carrier frequencies so that interference from other transmissions can be avoided.

Modulation is an essential part of any radio frequency transmitter circuit and is also a component of modems used for data transmission. In fact, the *mo* of the word *modem* stands for *mo*dulation and the remaining three letters stand for *dem*odulation.

This chapter discusses amplitude modulation and its offshoots. Amplitude modulation was the first type of modulation developed and is likely still the simplest and most common of all the techniques used today.

In this chapter we will look at:

1. What modulation does to a signal.

2. What frequencies are created.

3. The mathematics of amplitude modulation.

4. The circuitry.

5. The advantages and disadvantages of AM variations.

6. Demodulator circuits for all forms of amplitude modulation.

7. The performance of modulation in the presence of noise.

5.1 INTRODUCTION TO AMPLITUDE MODULATION

Any form of modulation must start with a high-frequency carrier signal that is a clean, stable sine wave. That is why we discussed oscillator circuits before look-

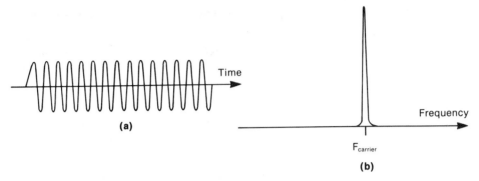

Figure 5.1 Oscilloscope (a) and spectrum analyzer (b) views of an unmodulated carrier signal.

ing at modulation. An unmodulated carrier would appear as shown in Figure 5.1. On an oscilloscope, you would see a constant amplitude and constant frequency sine wave. On a spectrum analyzer only a single[1] vertical line would appear.

Now to modulate the carrier. This can be done in any of three different ways. Its amplitude can be changed—this produces amplitude modulation (AM). Its frequency can be changed—frequency modulation (FM). Or, its phase can be changed—phase modulation (PM). In this chapter, we will start with amplitude modulation. But, first, consider this general rule for all modulation techniques.

> When a carrier signal is altered in any way, whether in amplitude, frequency, or phase, new signals called *sidebands* will appear. These will occur symmetrically above and below the carrier frequency and be spaced by an amount equal to the signal frequency that is doing the modulating. Some modulation techniques may even produce multiple sidebands, each spaced from the carrier by integer multiples of the modulating frequency.

Remember that this is a general rule for all modulation methods. Now we will look at amplitude modulation in detail.

[1] A single frequency with no jitter or noise should appear as a thin vertical line on a spectrum analyzer. However, when the student looks at the display on any analyzer he or she will see a thicker line, especially near the bottom. There are two reasons for this. First, the analyzer uses filters that scan over the selected frequency range to find the signals present. These filters, as mentioned in Chapter 3, can never have vertical sides and so the display is partially showing the response of these filters. The second reason is that no sine wave is perfect and will always have some noise and jitter. These problems show up as low-level sidebands close to the bottom of the vertical line. Therefore, the spectrum analyzer displays drawn in this text will show the lines as they are normally seen in practice.

Amplitude modulation (AM) is used in the 540 to 1600 kHz broadcast band, in the 27 MHz Citizens' Band, the aircraft band (108 to 136 MHz), and for numerous international shortwave broadcasts in the 3 to 30 MHz "shortwave" band. The picture portion of television transmission also uses amplitude modulation. (The color portion uses a combination of AM and phase modulation (PM).)

The modulation occurs when the carrier signal and the modulating signal (the music or voice) are fed into a nonlinear amplifier. This is shown in Figure 5.2 for the simple case where the modulating signal is a 2.0 kHz sine wave. The amplifier can't be any old high-distortion amplifier left over from your last lab project. It must have a square-law or second-order transfer characteristic. This was illustrated back in Figure 1.15. The amplifier output will have a number of different signals present and among them will be the original carrier at 1.00 MHz, the sum of the two input frequencies at 1.002 MHz called the *upper sideband*, and the difference at 0.998 MHz called the *lower sideband*. If these three are allowed to pass through a bandpass filter and all other signals are eliminated, the final amplitude modulated signal will emerge.

The output of the square-law amplifier before filtering contains all the frequency components shown in Figure 5.3. Only the three components near 1.000 MHz are the ones we want. If the bandpass filter has a bandwidth of at least 4.0 kHz (0.998 to 1.002 MHz), the carrier and its upper and lower sideband will pass through. The combination of these three produces the amplitude-modulated waveform. This is further illustrated in Figure 5.4.

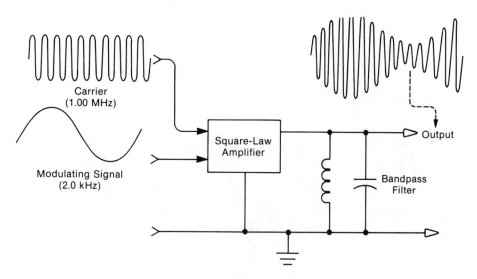

Figure 5.2 Generation of an amplitude modulated waveform using a nonlinear amplifier.

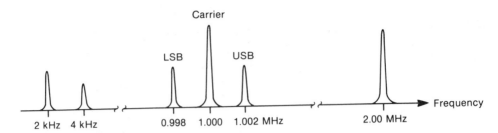

Figure 5.3 Frequency spectrum of the output of the square-law amplifier before filtering.

The original information is now completely described by the group of 1 MHz signals. The waveshape is visible as the "envelope" both at the top and the bottom of the modulated waveform. (Incidentally, there is no relation between the top and bottom of the waveform and "upper" and "lower" sidebands.) The amplitude of the original 2 kHz information is represented by the amplitude of each sideband—they are the same. The original frequency is the same as the distance from each sideband to the carrier.

Notice that each sideband contains the same information. You might be tempted to say that having two sidebands is a complete waste—in fact, many people unfortunately do. While there are advantages to eliminating one sideband, and these will be discussed later in this chapter, let's look at the advantages of keeping both first. With the same information carried at two frequencies, a certain amount of security is obtained and this is one of the benefits of modulation. As signals are transmitted over long distances, some deterioration of signal-to-noise ratio occurs through amplitude and phase distortion

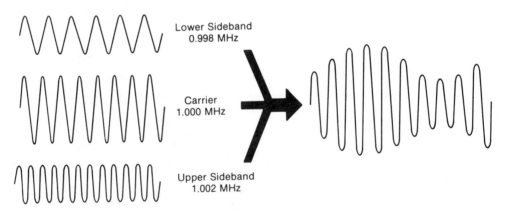

Figure 5.4 Three frequency components combine to form the amplitude-modulated waveform.

and added noise. The received signals will not be exactly the same as the transmitted ones and mistakes will be made in trying to interpret the information after demodulation. With the redundant information carried by two sidebands, it is like having the same story told to you twice (except for the boring parts); you can compare the two to see if they agree. The two sidebands are, therefore, a good thing. In fact, in the next chapter we will be examining frequency modulation (FM) which can produce an even greater number of sidebands and, therefore, greater redundancy and better noise immunity.

What limits are there to amplitude modulation? First, the carrier frequency must be at least twice the highest frequency in the modulating signal. This limit is shown in Figure 5.5(a). This happens to be a statement of what is called *Nyquist's Sampling Theorem*; you may run across this in some of your other courses. In this case, the tips of the carrier are actually describing periodic samples of the modulating signal. If the information is changing too fast, there won't be enough carrier peaks to fully describe it. The second limitation is the amplitude of the modulating signal. If it is too strong relative to the carrier strength, the carrier will be completely cut off for a while and the envelope won't look like the original signal; in other words, distortion will occur (and the

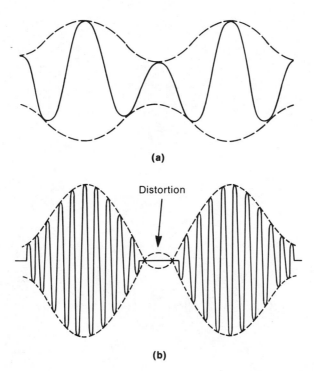

(a)

(b)

Figure 5.5 Two limiting conditions for amplitude modulation: (a) the lowest carrier frequency, and (b) the highest modulation.

Federal radio inspectors will be unhappy because the distortion generates unwanted signals outside of your assigned radio channel). This distortion is shown in Figure 5.5(b). The modulating signal should be just strong enough to cause the two sides of the envelope to meet for a brief moment. This condition results in what is called *100% modulation*. Anything more than this is called *overmodulation* and should be avoided.

What about more complex modulating signals? A voice signal or the waveform from a scanned television picture, for example, contain many frequencies. The carrier would now be modulated with a "band" of frequencies. This situation is illustrated in Figure 5.6. As expected, the original modulating signal has created two sidebands—one above and the other below the carrier frequency. The meaning of the word *sideband* should now be a bit more obvious since there is now a "band" of frequencies on either side of the carrier.

Notice that the lower sideband is a mirror image of the upper sideband. This happens because the frequency information from the modulating signal is now contained in the distance of each sideband component from the carrier. The higher the modulating frequency, the further away from the carrier the component lies.

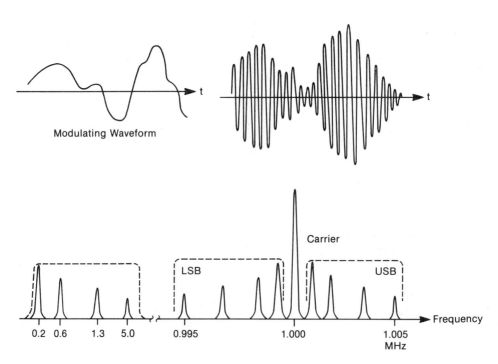

Figure 5.6 Modulation with a more complex signal produces a band of frequencies on either side of the carrier.

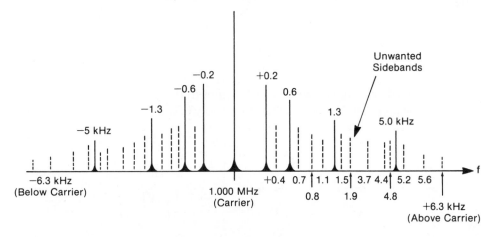

Figure 5.7 Unwanted signals are added to the upper and lower sidebands when nonlinear modulation occurs.

This is a good place to see why the modulating system had to use an amplifier with a square-law characteristic. If you follow the general rules from Chapter 1 for finding intermodulation components (page 19), you will see that the only signals that appear near the carrier after modulation are the ones from the original modulating signal. No new, unrelated signals appear. The envelope of the modulated waveform will look just like the original signal so no distortion will have occurred. Oddly enough, even though an amplifier with distortion has been used, the resulting signal, at least the part that we want, is not distorted. The result is called *linear modulation*.

Now we find out what happens if *nonlinear modulation* occurs. If you feel ambitious, find all the intermodulation products that could occur for the previous illustration if the modulator had a third-order characteristic in addition to the second-order part. These are found by considering all the sum and difference signals of three input frequencies at a time. These, along with the original signals and their third harmonics, will form a big list. Figure 5.7 shows the ones that lie close to the carrier frequency. The resulting envelope will no longer look like the original signal so distortion will have occurred.

5.2 MATHEMATICS OF AMPLITUDE MODULATION

Some further insight into amplitude modulation can be had through the use of mathematics. As an example we will consider a nonlinear amplifier with an input-to-output relation (transfer function) given by:

$$e_{out} = 10e_{in} + 3e_{in}^2$$

The amplifier will, therefore, have a linear gain component of 10 and a second-order or square-law gain of 3. The relationship is shown graphically in Figure 5.8.

Now let's place two sine waves at the input to the amplifier. The first is our modulating signal which is at a radian frequency of ω_m radians per second and

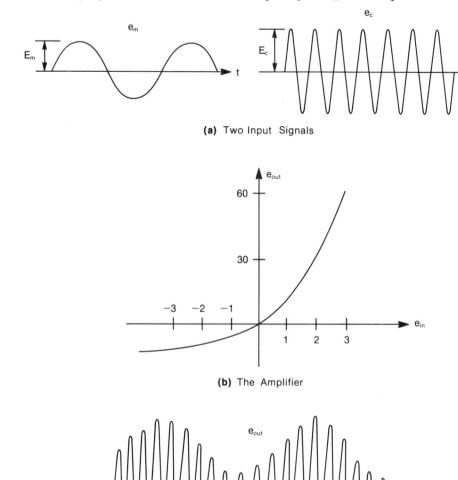

(a) Two Input Signals

(b) The Amplifier

(c) Output Signal

Figure 5.8 Two input signals (a), fed into an amplifier with a square-law gain characteristic (b), will produce the output waveform shown in (c). After filtering, this will be an amplitude-modulated wave.

has a peak amplitude of E_m volts. How this voltage changes with time is given by:

$$e_m = E_m \sin (\omega_m t)$$

The other input is the carrier at a radian frequency of ω_c radians per second. Its equation is:

$$e_c = E_c \sin (\omega_c t)$$

The output of this amplifier will be the waveform shown in Figure 5.8(c). It doesn't exactly look like the amplitude modulated waveform that we are after. However, if you look closely, you will see that the high-frequency portion of this wave is actually changing in amplitude but it also seems to be slowly riding up and down on a lower frequency signal. After passing through a bandpass filter to get rid of the unwanted components, the upper and lower envelopes of the waveform will be symmetrical. The result will be our AM waveform.

Now to put this into mathematical form. The input to the amplifier will be the two sine waves.

$$\begin{aligned} e_{in} &= e_m + e_c \\ &= E_m \sin (\omega_m t) + E_c \sin (\omega_c t) \end{aligned}$$

The output of the amplifier will be:

the carrier

$$e_{out} = \underbrace{10E_m\sin(\omega_m t) + 10E_c\sin(\omega_c t)}_{\text{linear amplification}} + \underbrace{3\big[E_m\sin(\omega_m t) + E_c\sin(\omega_c t)\big]^2}_{\text{square-law amplification}}$$

The first portion of this output is the same as the two input signals except that an amplification of ten times has occurred. The second portion is the square-law part and we will now use some trigonometric identities to expand this further. For clarity, we will temporarily leave the linear portion out.

$$\begin{aligned} e_{out} = \ ..... \ + 3\big[&\big(E_m\sin(\omega_m t)\big)^2 + \big(E_c\sin(\omega_c t)\big)^2 \\ &+ 2\big(E_m\sin(\omega_m t)\big)\big(E_c\sin(\omega_c t)\big)\big] \end{aligned}$$

The two identities that we will need are:

$$(\sin x)^2 = \frac{1}{2}(1 - \cos 2x)$$

$$2 \sin x \cdot \sin y = \cos (x - y) - \cos (x + y)$$

Using these, we can now expand the square-law portion of the output signal.

$$e_{out} = \dots + 3\left[\frac{E_m^2}{2}(1 - \cos 2\,\omega_m t) + \frac{E_c^2}{2}(1 - \cos 2\,\omega_c t)\right.$$

second harmonics

$$\left. + E_m \cdot E_c\left(\cos (\omega_c - \omega_m)\, t - \cos (\omega_c + \omega_m)t\right)\right]$$

lower sideband   upper sideband

This, along with the linear portion, is the full equation of the output waveform. The upper and lower sidebands were formed by the multiplication of two sine waves. The carrier came from the linear amplification portion of the equation. All of the other output parts are unwanted. They consist of the original modulating signal and its second harmonic, along with the second harmonic of the carrier. These three components can be easily removed with a bandpass filter that passes only the carrier and its upper and lower sidebands.

The relative amplitude of the carrier and the two sidebands at the amplifier's output must be carefully controlled. It is generally desirable to keep the upper and lower envelopes of the AM wave from touching each other, although, if the right techniques are used, they can actually cross. (Both the transmitter and receiver have to be especially designed to accommodate this.) It is still desirable to keep the envelopes as large as possible; otherwise some of the carrier potential is being wasted and the signal-to-noise ratio at the receiver end will suffer slightly.

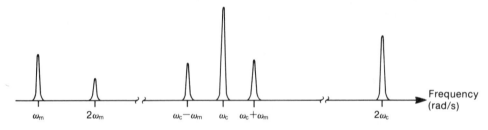

Figure 5.9 Output spectrum of the square-law amplifier.

When the top and bottom envelopes just touch, this is referred to as *100% modulation*. At this point, the voltage of each sideband (if a single modulating frequency is used) will be one-half the carrier voltage. This also means that each sideband will represent a signal power one-quarter that of the carrier when single-frequency, 100% modulation is used.

The relative amplitudes of the carrier and modulating signal that have to be fed "into" the square-law amplifier would depend on the amplifiers' exact gain equation. For the amplifier used in this section, a carrier input of 2 volts and a modulating input of 5/8 volt would result in a 100% modulated wave at the output.

For less than full modulation, the amplitude of each sideband would be:

$$E_{sideband} = \frac{m}{2} E_{carrier} \text{ (volts)}$$

where m represents the percentage modulation (i.e., m = 1.0 for 100% modulation)
$E_{sideband}$ is the amplitude of each sideband.
$E_{carrier}$ is the carrier amplitude after the amplifier.

The power in each sideband can also be calculated:

$$P_{sideband} = \frac{m^2}{4} P_{carrier} \text{ (watts)}$$

This is true only when a single sine wave is modulating the carrier because the shape of the modulating waveform determines its rms value and, therefore, its power. A square wave, for example, contains many frequencies and its rms value is 41% higher than a sine wave with the same peak amplitude.

The value of m can be found from the maximum and minimum voltage of the amplitude-modulated wave as viewed on an oscilloscope. This is shown in Figure 5.10.

Example 5.1 ━━━━━━━━━━━━━━━━━━━━━━━━━━━━━━━━━━━━

This example consists of a sequence of questions and answers pertaining to amplitude modulation. The questions all refer to the same carrier and its sidebands.

> 1. If the carrier frequency of an AM wave is at 2.5 MHz and it is modulated with a 5 kHz sine wave, what are the sideband frequencies?

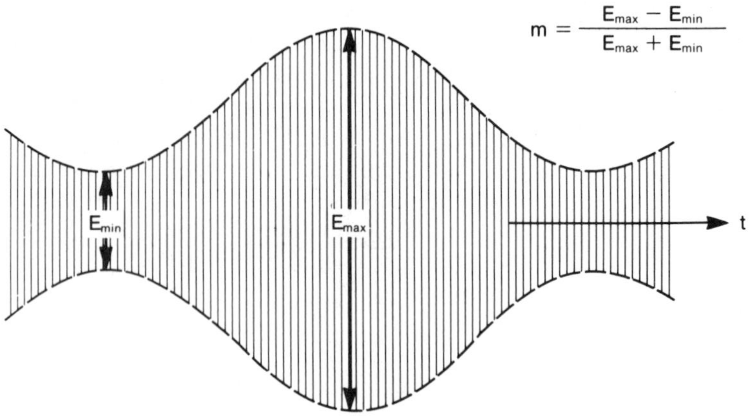

$$m = \frac{E_{max} - E_{min}}{E_{max} + E_{min}}$$

Figure 5.10 Percentage modulation is determined from the minimum and maximum amplitudes of the AM waveform.

The sidebands would appear 5 kHz above and below the carrier frequency. Therefore, the sideband frequencies would be:

$$= 2.500 \text{ MHz} \pm 5.0 \text{ kHz}$$

$$= 2.495 \text{ and } 2.505 \text{ MHz}$$

2. If the carrier's amplitude is 10 volts rms and it is modulated to 50% ($m = 0.5$) by the modulating signal, what will the amplitude of each sideband be?

Each sideband will have an rms amplitude of:

$$E_{sideband} = \frac{m}{2} E_{carrier}$$

$$= \frac{0.5}{2} \times 10$$

$$= 2.5 \text{ volts rms}$$

3. If the carrier power is 2.0 watts (10 volts across a 50 ohm resistor) how much power is contained in each sideband?

Each sideband will have a power of:

$$P_{sideband} = \frac{m^2}{4} P_{carrier}$$

$$= \frac{(0.5)^2}{4} \times 2.0 = 0.125 \text{ watts}$$

This, incidentally, brings the total power of the AM wave up to $2 + 0.125 + 0.125 = 2.25$ watts.

4. With reference to Figure 5.10, what would the minimum and maximum voltages be for this AM wave if it was looked at on an oscilloscope.

Figure 5.4 can also be used to help with this question. It shows that the three individual sine waves combine to form the AM wave. We know the rms amplitudes of all three signals but, for this question, we will have to work with peak-to-peak values.

$$\text{Carrier} \qquad = 10 \times 2\sqrt{2} = 28.28 \text{ volts p-p}$$

$$\text{Each sideband} = 2.5 \times 2\sqrt{2} = 7.07 \text{ volts p-p}$$

The largest amplitude occurs when all three sine waves are momentarily in phase.

$$E_{max} = 28.28 + 7.07 + 7.07$$

$$= 42.42 \text{ volts p-p}$$

The smallest amplitude occurs when both sidebands momentarily have the same phase but the carrier has the opposite phase.

$$E_{min} = 28.28 - (7.07 + 7.07)$$

$$= 14.14 \text{ volts p-p}$$

Just for a check, we should be able to use these two values to get the original value of $m = 0.5$

$$m = \frac{E_{max} - E_{min}}{E_{max} + E_{min}}$$

$$= \frac{42.42 - 14.14}{42.42 + 14.14}$$

$$= 0.5 \quad \text{(fortunately)}$$

5.3 AMPLITUDE MODULATION CIRCUITS

Now it is time to look at some actual circuits that can be used to amplitude modulate a carrier with a lower frequency modulating signal. Two circuits will be described, each showing a different approach.

We have indicated that the process requires a non-linear device with a second-order gain characteristic. This isn't too hard to achieve as the first circuit shows.

5.3.1 Diode Modulator

A very simple amplitude-modulator circuit is shown in Figure 5.11. The diode is the nonlinear device and will approximate the required second-order characteristic as long as the two input signals are kept large. This makes the forward drop of the diode and its changing forward resistance less significant.

The two input signals are added together through the two resistors R_1 and R_2. This is the same type of linear mixing that occurs in audio system microphone "mixers." No new frequencies are generated yet. The waveform at this point in the circuit (before the diode) shows the high-frequency carrier signal to be riding up and down on the lower-frequency modulating signal.

Next, the combination is rectified by the diode so that only the top half passes through. This is a nonlinear process and all the new frequencies appear. It is important that the carrier voltage is stronger than the modulating signal voltage before the combination gets to the diode. Otherwise, the diode will not pass the full upper envelope; its lower portion will be clipped off. The bandpass filter then allows the carrier and its newly formed sidebands to pass through while eliminating the original modulating signal and all unwanted harmonics.

5.3.2 Class-C Transistor Modulator

While the previous circuit will work very well, it doesn't provide any gain. In fact, after passing through the two summing resistors (R_1 and R_2) and then having the bottom chopped off by the diode, there is a considerable overall loss. The next circuit is capable of operation at high-power levels with good efficiency and can also provide some gain. It is often referred to as a *high-level modulator* because of the relatively high power the modulating source must have.

The circuit is shown in Figure 5.12. The base of the transistor is biased at ground potential with resistor R_1. Therefore, the transistor will be normally cut off. A strong carrier signal applied to the base will turn the transistor on every time the voltage swings above + 0.6 volts and then turns it off again when the input swings below 0.6 volts. This defines the transistor's operating conditions

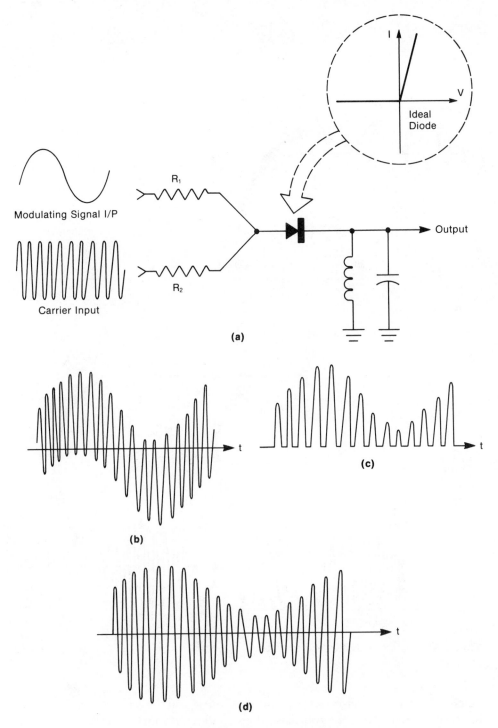

Figure 5.11 Diode modulator, (a). Before the diode, two signals are added together (b), the diode then half-wave rectifies the combination (c) and, after bandpass filtering, the final AM wave appears (d).

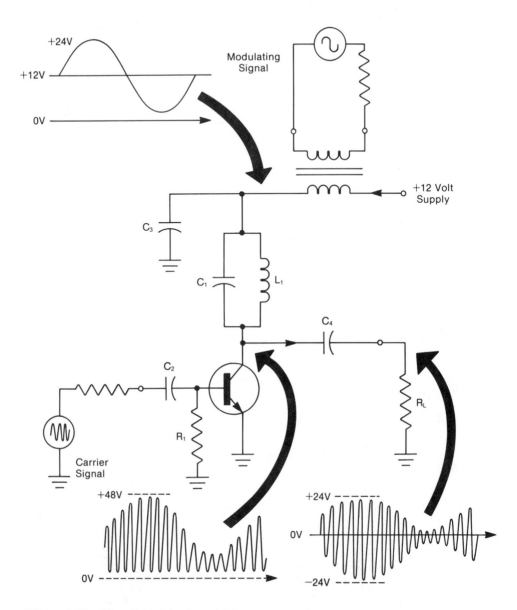

Figure 5.12 Class-C high level modulator.

as class-C since collector current will flow for less than one-half cycle of the input wave. Each time the transistor does conduct, it must be driven right into saturation and the designer must be sure that this happens for all possible values of the collector supply voltage.

The basic supply voltage is shown as 12 volts but the modulating voltage is added in series with it before it reaches the transistor. This addition is done with a transformer so that the total voltage will swing both above and below the 12 volt value. As mentioned, the modulating signal must come from a reasonably powerful source since it, along with the main supply, is feeding power into the transistor. The effective power supply can, therefore, swing up to a high of $+24$ volts and right down to 0 volts. As long as the carrier is strong enough to keep the transistor going in and out of saturation even when the collector voltage is at its peak value, linear modulation will occur.

The waveforms that accompany Figure 5.12 help to explain how the circuit works. A tuned circuit formed by C_1 and L_1 is resonant at the carrier frequency and will have a bandwidth wide enough to pass the carrier and its new sidebands. As the effective supply voltage swings between 0 and $+24$ volts, the carrier continually drives the transistor between saturation and cut off. The collector waveform is, therefore, a high-frequency sine wave whose lower tips always stop at 0 volts (transistor in saturation) and whose upper tips rise to twice the instantaneous supply plus modulating voltage. This doubling is due to the energy stored in the tuned circuit during transistor saturation and released, at the resonant frequency, when the transistor is cut off. Notice that the collector voltage actually rises to four times the basic power supply voltage. This is important if you plan to stick your finger into a transmitter someday or, even if you decide to design one, for both your finger and the transistor must be capable of standing the high voltages present.

Capacitor C_4 along with R_L form a high-pass filter. This removes the low-frequency modulating signal that is part of the collector waveform leaving the symmetrical modulated waveform. The load resistor could represent the input resistance of a following amplifier stage or it could be the "radiation resistance" of a transmitting antenna.

Capacitor C_3 is a power supply filter. It is just large enough to stop the high-frequency carrier from getting into the power supply but not large enough to stop the modulating signal from reaching the collector of the transistor.

A look at the flow of power through this circuit is interesting. This is illustrated in Figure 5.13. Remember that a transistor does not create energy; all it can do is control the flow of power from the supply. In the case of this circuit, the "supply" is made up of two parts—the basic 12 volt supply plus the modulating signal source.

Assume, for example, that when a carrier signal alone is applied to the transistor and no modulating signal comes through the transformer, the transistor draws 10 watts from the power supply (12 volts @ 833 mA). The output signal that reaches R_L will contain only a carrier signal and it will be less than the DC input power because the transistor's efficiency will be less than 100%. A typical efficiency for this type of circuit is 70%. Therefore, the output carrier power would be 7.0 watts. Now, as the modulating signal (the audio) is increased in amplitude, sidebands will appear at the output along with the carrier

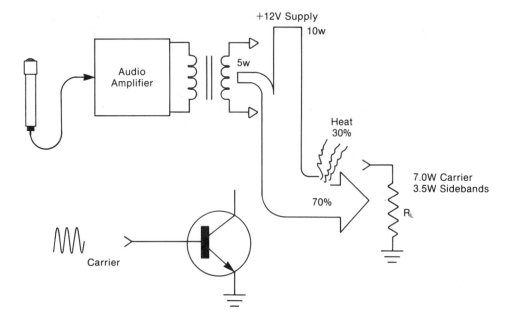

Figure 5.13 Flow of power in the class-C modulator.

signal. The power of the carrier will not change; rather, the total output will increase due to the added sideband power. This extra power comes as the collector supply voltage swings above and below the 12 volt value and it is the power from the audio amplifier that is doing this. The audio amplifier itself is also drawing power from the power supply. As with the carrier, 70% of the power coming into the transistor from the modulating transformer will reach the output—in this case, as extra sideband power.

At 100% modulation, we have already seen that the power of the two sidebands will be one-half of the carrier power. Therefore, for single sine wave modulation, the audio amplifier must provide 5.0 watts to the modulating transformer; 70% of this will reach the load for a total power in sidebands of 3.5 watts. The total input power to the transistor from both the basic supply and the audio amplifier will be 15 watts. The total output to the load resistor will be 10.5 watts (carrier plus sidebands). The remaining 30% or 4.5 watts is unfortunately wasted as heat that keeps the transistor warm.

5.3.3 Digital Amplitude Modulation

A third example of an amplitude modulator circuit is made with the digital logic gates. This could be used for the transmitter portion of a modem for data communications. Any of the basic logic gates (AND, OR, NAND, NOR) could be used; the example in Figure 5.14(a) happens to use an AND gate. This gate's output will be Hi only when both inputs are Hi at the same time. The output contains a 100% modulated wave plus a portion of the original data signal as well as harmonics of both input signals (because the signals are all rectangular waveforms). Bandpass filtering could be added, if desired, to pass only the carrier and its sidebands.

In some cases, less than 100% modulation is desired. This leaves a sample of the carrier signal present at all times which might be useful at the receiving end. The circuit of 5.14(b) adds two resistors to the original circuit and this will add about 10% of the input carrier signal to that at the output. Remember that the result, in this case, now has three distinct voltage levels and so is no longer a digital signal. Any further amplification requires linear amplifiers.

Data communications and modems will be further discussed in Chapter 11.

5.4 AM AND ROTATING VECTORS

A handy way of describing an AM waveform is to consider each of the component signals as voltage vectors that add up to produce one resultant voltage vector. The student should already be familiar with the vector form of a single sine wave. As a review, this is shown in Figure 5.15.

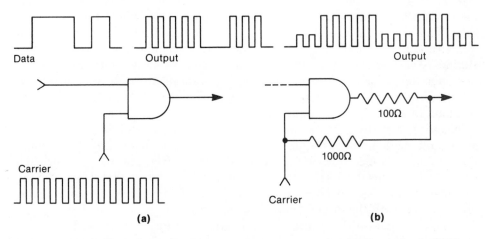

Figure 5.14 Digital data can amplitude modulate a high speed carrier. A simple AND gate (a) results in an output with 100% modulation. If a lower value of modulation is needed, the circuit of (b) can be used.

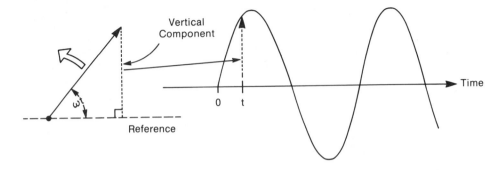

Figure 5.15 A single rotating vector (a) describes a sine wave (b). Its amplitude at any instant in time is equal to the vertical component of the rotating vector.

For amplitude modulation, consider the simplest case of one lower sideband frequency, an equal upper sideband, and a carrier. The lower-side frequency can be represented by a vector rotating at the same speed as the lower sideband frequency, the carrier by a longer vector rotating at a slightly faster rate, and the upper side frequency by an even faster rotating vector. Now, trying to visualize three vectors of two different lengths rotating at three different rates is a bit difficult. Therefore, for a start, let the carrier vector remain at rest in the vertical direction for a while. The other two will now move relative to this stationary vector. The lower-side frequency vector will now rotate slowly in the one direction and the upper-side frequency will rotate at the same slow (audio) rate but in the other direction. (Its frequency is higher than the carrier's.) The resultant of these three vectors describes the envelope of the AM wave. As the sideband vectors rotate at the modulating frequency rate and the carrier vector stays motionless, no high frequency component is described, only the modulation envelope. Notice how the resultant always stays in line with the original carrier vector. This means that the high-frequency structure that appears in the modulated wave (as viewed on an oscilloscope) remains at the same frequency and in phase with the original carrier sine wave.

Figure 5.16 also shows the resultant when all three vectors are allowed to rotate. Now they all turn at their higher frequency rates and so the full AM wave appears. This vector representation will be used to help with other forms of modulation and will help when comparisons are made.

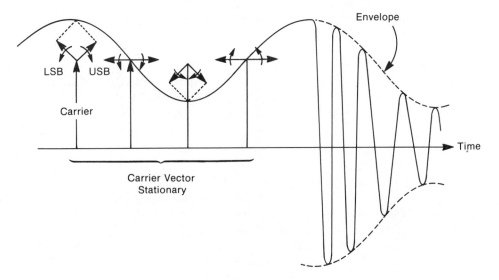

Figure 5.16 Three rotating vectors describe the simplest AM wave. When the carrier vector remains vertical and the sidebands rotate at their audio rate, the envelope is described. When all vectors rotate at their normal speeds, the full AM wave is obtained.

5.5 DOUBLE SIDEBAND SUPPRESSED CARRIER

Now we will examine a variation of full amplitude modulation. We have seen that even with full modulation (100%), the carrier represents about two-thirds of the total transmitted power (depending on the modulating wave shape). While the carrier signal was a necessary part of the modulation process, it has no value during the actual transmission (via cable or radio wave) of the modulated signals. The sidebands now exist at high frequencies close to that of the original carrier and are perfectly capable of traveling by themselves. The word *carrier* unfortunately tends to suggest that it is the essential transportation mechanism—it isn't. The carrier, however, does have a function at the receiving end where the demodulation occurs. Here it is used as the reference frequency to determine what modulating frequency each sideband signal represents. For example, a sideband at 1.005 MHz will represent a 5000 Hz modulating signal only when it is compared to a carrier sine wave at 1.000 MHz. If a different carrier at 1.001 MHz were accidentally reinserted at the receiver, the sideband would be mistakenly interpreted as representing a 4000 Hz audio tone. The carrier signal does, therefore, have a purpose but a normal AM transmitter pays a heavy penalty just to include a "reference signal." The transmitter needs to produce three times the power output of a comparable station that doesn't include the carrier. For commercial broadcasting, this penalty is tolerated since the radio receivers are much simpler and are easier to tune.

For greater transmitter efficiency, the carrier can either be reduced in strength or completely eliminated. A popular example is the *difference channel* used to carry part of the stereo sounds over commercial FM channels (88 to 108 MHz). Such a signal with the carrier removed is referred to as a *double side-band suppressed carrier* (DSBSC) signal.

The differences between the ordinary AM wave and the more efficient DSB suppressed carrier wave are shown in Figure 5.17. The time domain waveform for the AM signal shows two totally separate 5 kHz envelopes—one at the top and the other at the bottom of the wave. The high-frequency part in the middle is the original carrier frequency. The DSB suppressed carrier waveform also has two 5 kHz envelopes but they aren't separate; they overlap. If you look closely at the drawing, you will also see that the high-frequency pattern in the middle keeps reversing its phase relative to that of the original carrier (shown in the drawing just for reference). The phase reversals correspond to the movement of the modulating signal envelope as it changes from the positive half to the negative half. The correct envelope is still there but the detector circuit in the receiver is going to have to keep track of which half of the wave it's supposed to be using. In other words, the detector will have to be "phase sensitive," not just amplitude sensitive, and this increases its cost.

The DSB suppressed carrier wave with the same sideband strength is also half the peak-to-peak amplitude of the corresponding AM wave (10 Vp-p compared to 20 Vp-p). This results in a two-thirds reduction in the transmitter power output needed for the DSBSC wave (4.77 dB reduction). (The 20 volt to 10 volt reduction is not a 6.02 dB saving because the two waves are not the same shape and so their rms voltages do not have the same ratio as their peak-to-peak voltages.)

Notice also that the suppressed carrier signal has twice as many "bumps" on its top side and its bottom side, as does the AM wave. If this signal is listened to on an ordinary AM receiver, the "sound" will be very difficult to understand. The audio from the loudspeaker will have unnatural high-frequency components. Use of a proper detector will solve the problem.

5.5.1 Suppressed Carrier Modulators

The obvious way to remove the carrier from the AM signal is with a good band rejection filter. Unfortunately, if a "Hi-Fi" music signal (20 to 20,000 Hz) were modulating a 1.000 MHz carrier then the lower sidebands would lie just 20 Hz below the carrier and the upper sidebands would begin just 20 Hz above the carrier frequency. The filter would, therefore, have only a 40 Hz bandwidth in which to work and would need a Q of much more than $\frac{1.000 \text{ MHz}}{40 \text{ Hz}} = 25,000$. This makes that idea rather impractical.

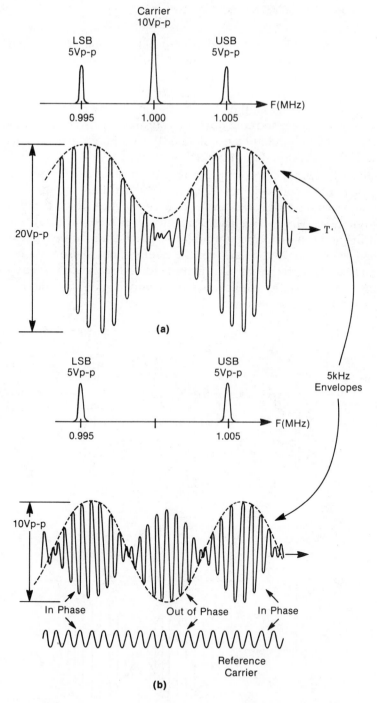

Figure 5.17 Comparison of an AM wave (a) and a DSB suppressed carrier wave (b) both in time and in their frequency spectra. The modulating frequency and amplitudes are the same in both cases. Note the reduced overall amplitude of the DSB waveform.

The preferred method is to use circuits called *balanced modulators*. The word *balanced* means that one or more of the input signals (carrier, audio) will not appear as part of the output yet will still contribute to the modulation process. One of these circuits is the *ring modulator* shown in Figure 5.18. The word *ring* describes the fact that all the diodes point in the same direction around a circle. (This makes it easy to remember for your next test.) The ring modulator is a double-balanced modulator since neither the audio or the carrier appear at the output. This makes the filtering much easier.

The four diodes that make up the ring do not act as rectifiers. Instead, they form pairs of on/off switches. The carrier signal must be much stronger than the audio signal and it is the carrier that determines which two diodes are on at any one time. The weaker audio signal will then follow a low resistance path from the audio transformer through to the RF transformer. Because it is weaker than the carrier, the polarity of the audio signal doesn't matter. Audio current can flow through the on diode pair in either direction. This may sound a little strange. Actually, the audio current simply makes the diode current stronger or weaker but never reversed.

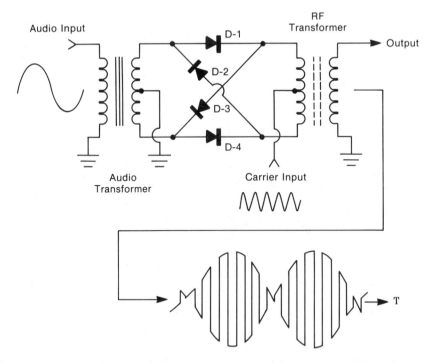

Figure 5.18 Ring modulator provides double sideband suppressed carrier modulation. It is a double-balanced modulator since neither the audio nor the carrier appear at the output.

When the carrier input terminal is more negative than the ground terminal, the primary of the RF transformer will be more negative than the secondary of the audio transformer. The two horizontal diodes, *D*-1 and *D*-4, will, therefore, be forward-biased and so will create a path for the audio to reach the RF transformer. This connection only lasts for one-half cycle of the carrier. On the other half-cycle, the diagonal pair of diodes will conduct and again the audio will be connected to the RF transformer, this time through a crossed connection. This produces a short negative peak at the output even though the input audio is still on its positive half-cycle. The resulting output waveform is, therefore, a continual sequence of inverted and noninverted samples of the audio waveform. The output wave shown in Figure 5.18 has sharp edges to emphasize the fact that the diodes turn on and off abruptly and act as switches rather than rectifiers. A bandpass or low-pass filter would easily get rid of the resulting harmonics.

A second balanced mixer circuit is shown in Figure 5.19. It is also a double-balanced design and is available in integrated circuit form. Its big advan-

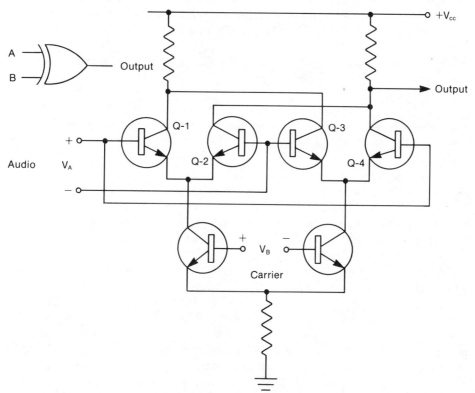

Figure 5.19 Integrated circuit version of a double-balanced modulator. Two differential amplifiers share common load resistors. The two amplifiers are connected with opposite outputs. The carrier input at *B* determines which amplifier is on and which is off.

tage over the previous circuit is that it doesn't need transformers. The circuit is actually the analog equivalent of a logic Exclusive-OR gate.

The circuit consists of two differential amplifiers sharing one pair of load resistors. The collectors of the second amplifier (Q-3 and Q-4) have opposite connections to the load resistors than the first amplifier (Q-1 and Q-2). The carrier signal at the B input terminal determines which differential amplifier is working and which is off. The + and − marks on the input terminals are the same as those used on operational amplifier input terminals to indicate noninverting and inverting inputs, respectively. These inputs are *differential inputs* which means that both terminals will have changing voltages on them. One input terminal will be going more positive when the other is going more negative by the same amount. The inputs that you are likely more familiar with are unbalanced connections using an active wire and a ground wire.

The modulator operation can be described as follows. If the B + terminal is more positive and the B − terminal is less positive then the left differential amplifier (Q-1 and Q-2) will be operating. Under this condition, an input audio signal at A+ will appear at the left load resistor. One-half cycle of the carrier later, the B− input will be more positive and the right differential amplifier will be operating (Q-3 and Q-4). Now an input signal at A + will end up at the right load resistor which is the opposite of the previous condition. The transistor circuit is, therefore, operating the same as the previous diode ring modulator and the output waveforms will be the same.

5.6 SINGLE SIDEBAND MODULATION

The next possible step after removing the carrier from the AM wave is to eliminate one of the sidebands. The result is called *single sideband* (SSB). Remember that having both sidebands isn't a complete waste. The redundancy actually reduces the effects of noise and interference on the modulated signal. However, the big disadvantage of the double sideband signals is that they use up twice the bandwidth that the original modulating signal did. In the same frequency band, for example, where CB can hold only 40 AM channels, the use of SSB allows 80 to be squeezed in instead.

Single sideband is extensively used by CB operators, radio amateurs, for military communications, and for telephone trunk lines (lines between switching centers, not to your homes). In all cases, the reduced bandwidth is the important feature. SSB is rarely used where any high-quality music must be transmitted; most of its use is for limited bandwidth voice communications.

The block diagram of an SSB modulator is shown in Figure 5.20. This is the easiest and most common method. A balanced modulator is fed with the audio signal and a carrier (1.00 MHz in this example). The output of the modulator will have both the upper and the lower sidebands but no carrier. A

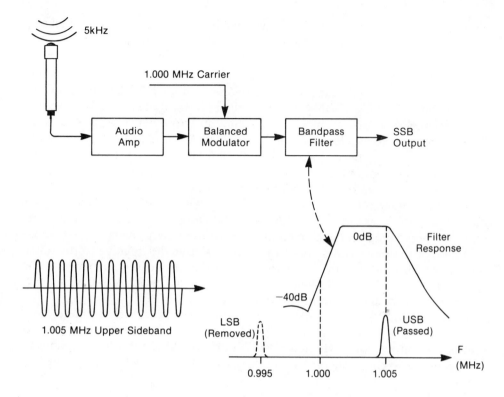

Figure 5.20 Single sideband signals can be generated with a balanced modulator and a bandpass filter. A typical output is shown for a single 5 kHz audio input.

bandpass filter then removes the unwanted sideband—either one. This is still a moderately difficult filtering job but one that can be handled with high-Q crystal filters. If only voice signals are handled, only audio frequencies above about 300 Hz are important. The bandpass filter, therefore, has a 600 Hz (0.9997 to 1.0003 MHz) range in which to increase its attenuation from nearly 0 to 40 dB or more. The frequency response of the necessary filter is included in Figure 5.20. This is the *filter method* of generating SSB signals. A second method will be described shortly.

Figure 5.20 also describes the output signal from the SSB modulator. If a person whistles into the microphone, making a perfect 5 kHz sine wave, the only output signal will be at 1.005 MHz. On an oscilloscope, this single high-frequency sine wave appears. It has no amplitude variations showing (as long as the person whistling can hold out). On a spectrum analyzer, only a single vertical line will appear.

For testing purposes, something more exciting than a single audio tone is needed. Two simultaneous audio tones are often used—600 and 1000 Hz are common. Two high-frequency sine waves will be generated in the one sideband and create the pattern with the amplitude variation shown in Figure 5.21. This pattern now allows the "linearity" of the modulator to be checked. Each envelope "should" look like a perfect sine wave. If they don't, something is wrong.

It should be noted that the two envelopes will be 400 Hz sine waves not 600 or 1000 Hz. This is because the SSB signal has only two frequencies in it and they are 400 Hz apart. The high-frequency structure between the two envelopes, as seen on an oscilloscope, will not be 1.000 MHz. Instead, it will be some varying frequency between 1.0006 and 1.0010 MHz, depending on the relative strengths of the two tones.

The second method of creating SSB signals is shown in Figure 5.22. It does away with the selective filter but replaces it with some fairly elaborate and difficult to adjust circuitry. Two balanced modulators are used and each is fed with the modulating audio signal and the high-frequency carrier signal. However, one of the modulators gets all its inputs shifted 90° relative to the other modulator. This phase shift is easy to create at the carrier frequency because it is a single-frequency signal. The phase shift over a bandwidth of 200 Hz to 5000 Hz for the audio signal is quite a different matter.

Both balanced modulators create double sideband signals without carriers. The upper sidebands of each are shifted 90° in the same direction and so add at the output point. The lower sidebands are shifted 90° in the opposite direction and so they cancel. The result is an SSB signal with the upper sideband only.

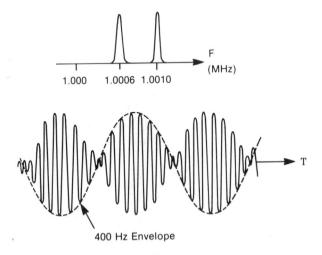

Figure 5.21   SSB modulator output during a two-tone test. The carrier was at 1.000 MHz and the two audio tones were 600 and 1000 Hz. The resulting envelopes will vary at a 400 Hz rate.

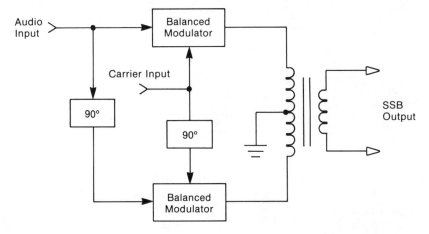

Figure 5.22 SSB generation using phase shifts to cancel one of the sidebands. The 90° shift for the audio signal is the most difficult to create properly.

Now for the disadvantage of SSB. If you look at the frequency spectrum that is part of Figure 5.21 you will notice that there are two frequencies in the upper sideband. What audio frequencies do they represent? They are supposed to represent 600 and 1000 Hz tones but how is the receiving system supposed to know this? This is the big problem. A carrier signal is needed at the receiver for demodulation but there is nothing in the transmitted signal to indicate what its exact frequency should be. When SSB stations are tuned in on a shortwave or CB receiver, the operator has to adjust the tuning (clarifier control on CB) for the most natural sounding voice frequencies. Even the telephone trunk line system can shift voice frequencies by as much as 10 Hz (not really noticeable). Such performance is entirely unsuitable for Hi-Fi/stereo music and so SSB is usually restricted to voice-only operation.

5.7 AM WAVE DEMODULATION

The process that is needed at the receiver to take a modulated waveform and turn it back into the original voice or music or data is called *demodulation* or *detection*. We will start with a simple demodulator for the full AM wave and then look at SSB and DSB demodulators.

There is a very simple way to demodulate AM. Since either the upper or the lower modulation envelope has the same shape as the original modulating signal, a simple rectifier circuit with a low-pass filter is all that is needed. This is shown in Figure 5.23.

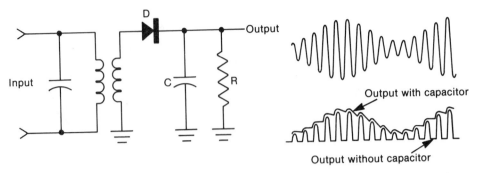

Figure 5.23 Simple AM demodulator or detector circuit. The RC time constant must be kept small enough so that the capacitor voltage can follow the negative slopes of the modulation.

When the modulated input signal to this circuit goes positive, the diode conducts and rapidly charges the capacitor up to the peak input voltage. Then, when the input starts to fall again, the diode turns off and the capacitor discharges slowly through its parallel resistor. This discharge time is critical because, if it is too long, the capacitor voltage will not be able to follow the modulation envelope on any rapid downward swings it might make. Such a problem is referred to as *diagonal clipping*. A resistor value of 10 kΩ and a capacitor value of .005 μF are typical.

Although this demodulator is simple, it isn't the best. It works well with strong, noise-free signals but when signals get weak and random noise becomes more of a problem, performance drops. There are two reasons for this. First, since the diode is a nonlinear device, intermodulation occurs between the noise and the desired signal, creating even more noise. As a result, the signal-to-noise ratio after the demodulator is worse than it was before. The second problem is that a simple diode doesn't really take advantage of the information redundancy in the two sidebands. The ideal AM demodulator shown in Figure 5.24 corrects both of these deficiencies.

The ideal AM demodulator is actually a balanced mixer. (A mixer and a modulator are the same thing.) This one happens to be a simpler version of the four-diode ring modulator shown previously. The circuit needs two inputs—one is the modulated AM wave and the other is a strong sine wave identical in frequency and phase to the original carrier. The diodes again work as switches, not as rectifiers, as was the situation in the simpler demodulator. When the strong local carrier goes positive, diode *D*-1 will be almost a short circuit. It acts as a switch to connect the input signal through to the capacitor for one-half cycle. Then, when the carrier goes negative, diode *D*-2 acts as a closed switch. But, it connects the opposite end of the transformer secondary to the capacitor. Therefore, even though the modulated input has swung down on its negative half-cycle, the diode is connecting another positive voltage to the capacitor. The

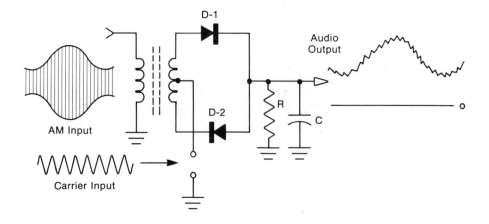

Figure 5.24 The ideal AM demodulator consists of a balanced mixer (modulator) and a low-pass filter. A strong sine wave identical to the original carrier frequency is needed.

circuit acts as a full wave rectifier in synchronism with the input signal. The capacitor acts as a filter to fill in the gaps between the peaks.

Another way of looking at this circuit's operation is to consider it as a mixer. When two signals are fed into a mixer, the output will contain the sum and difference of these inputs. For a double sideband input with a carrier in the middle, the sidebands are symmetrical about the carrier frequency. The two sidebands, for example, that are 2 kHz above and 2 kHz below the carrier will produce identical outputs from the mixer—namely 2 kHz. This output will be twice as strong as any frequency that appeared in only one sideband and not the other. This is illustrated in Figure 5.25. Each sideband has three identical signals. There is also an unwanted signal in the upper sideband at 1.5 kHz without a corresponding one in the lower sideband. Similarly, there is an unwanted signal in the lower sideband at 2.5 kHz. Notice the relative strengths of all the signals before demodulation and after. The unwanted signals are still there but are relatively weak. This is the advantage obtained by transmitting both sidebands and recovering their information with a good demodulator.

5.8 DSB AND SSB DEMODULATORS

Now we must develop some elaborate circuits to demodulate each of the remaining two variations of amplitude modulation. Actually, this has already been done. The balanced mixer circuit of Figure 5.24 will work on all types of amplitude modulation including double sideband suppressed carrier and single

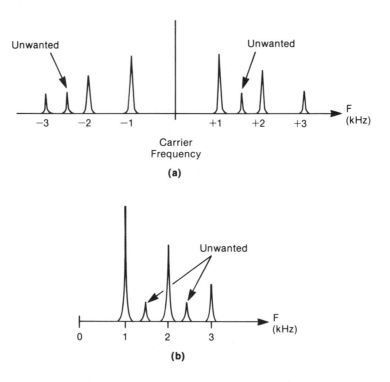

Figure 5.25 Before demodulation (a) a modulated wave has three good sidebands placed symmetrically about the carrier plus two unwanted noise frequencies. After demodulation (b) in an ideal demodulator, the noise is still there but is weaker in comparison.

sideband. The only problem is supplying a local carrier signal at a frequency and phase identical to that used in the original modulation process. With ordinary AM, this isn't a big problem because the carrier comes along with the signal. It is then a simple problem to synchronize a local oscillator with the incoming signal.

With DSB, there is no carrier. However, the sidebands are symmetrical about the frequency where the carrier originally was. This fact can be used to correctly recreate the local oscillator frequency. If this oscillator should drift, the results on the demodulated audio are rather bad. With a frequency drift, the oscillator will be closer to the one sideband and so all of its signals will sound low in frequency. At the same time, the oscillator will have moved further away from the signals in the other sideband and so all those frequencies will sound slightly higher. FM stereo transmission in the 88 to 108 MHz band uses DSB for part of the signal. To get away from frequency drift problems with oscillators at the receiver, a weak pilot carrier at 19 kHz is included at the transmitter end. At the receiver, this is doubled to 38 kHz which is the correct frequency and phase needed to demodulate the "difference channel."

SSB remains with the biggest carrier problem. There is just no way of knowing what frequency the carrier should be (unless a weak carrier is included). Normally, it takes a human operator to keep adjusting the carrier injection oscillator for the most natural sounding voice. However, for low-frequency SSB voice channel multiplexing, as used by the telephone companies, stable crystal oscillators can keep the voice frequencies correct to within 10 Hz.

5.9 AMPLITUDE MODULATION AND NOISE

In Chapter 1 we discussed noise and how it deteriorated signals as they travel between the transmitter and the receiver. Now that we have examined three variations of amplitude modulation, let us see how well they survive when noise is added to the signal. The situation we are discussing is shown in Figure 5.26. The modulated signal to be tested is fed into a demodulator through a bandpass filter. This filter is just wide enough to pass all the RF signals and no wider. Noise is also fed through the same filter and into the demodulator. The amount of noise that gets into the demodulator will depend on the bandwidth of the filter. For AM and DSB, the filter will have to be about 10 kHz wide (for normal voice modulation). For SSB, the filter will only be 5 kHz wide.

After the demodulator, the resulting "audio" is again filtered to pass only the significant modulating frequencies. In this case, the low-pass filter would cut off at 5 kHz.

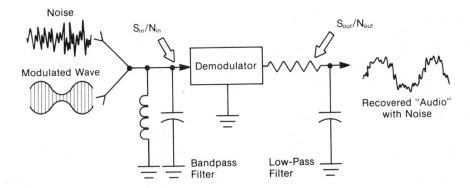

Figure 5.26 The performance of modulation techniques and demodulator circuits can be measured by comparing the input signal-to-noise ratio with the output signal-to-noise ratio.

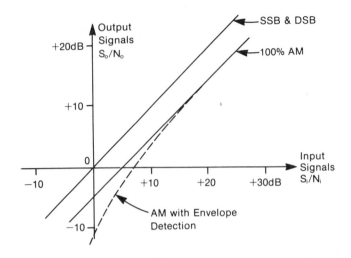

Figure 5.27 Comparison of input S/N ratios before detection with S/N ratios after detection. SSB and DSB using synchronous detection work very well. AM, with its carrier and using a simple envelope detector, works very poorly.

Figure 5.27 compares the signal-to-noise ratios of the input signals at the RF frequencies to the signal-to-noise ratios of the output signals at the audio frequencies. Both SSB and DSB perform equally well. Using a synchronous detector, these two modulation systems will have the same S/N ratio after demodulation as before. Remember that DSB requires twice the bandwidth that SSB does yet the performance is identical.

However, the AM signal suffers badly. Its S/N ratio after demodulation is 4.77 dB poorer than before demodulation. This is because the strong carrier forms two-thirds of the signal power before demodulation yet contributes nothing to the information afterwards. And this isn't all the problems that AM suffers from. The curve just mentioned assumed that a synchronous or "ideal detector" was used. If the ordinary diode envelope detector is used instead, the performance is about the same for strong signals but for weak signals the output S/N ratio becomes much worse.

5.10 REVIEW QUESTIONS

1. A carrier frequency of 830 kHz is amplitude modulated with a 5.2 kHz sine wave. Find the resulting sideband frequencies and the total bandwidth of the signal.

2. What type of modulation is used by each of the following? Radio stations between 540 and 1600 kHz? Radio stations between 88 and 108 MHz? Citizen's band? Television video from Earth transmitters? Television sound from Earth stations? The color portion of television? Television from satellites?

3. An amplifier has a nonlinear characteristic given by $E_{out} = 5E_{in} + 2E_{in}^2$. Two sine waves are fed to the amplifier's input: a 5 volt signal at 25 kHz and a 2 volt signal at 1.3 kHz. Calculate the amplitude and frequency of all output components. What is the resulting percent modulation?

4. An amplitude-modulated waveform is viewed on an oscilloscope. Its greatest amplitude is 9.3 volts p-p and its minimum amplitude is 3.5 volts p-p. What is the percent modulation? (See Figure 5.10.)

5. Do you think that three individual oscillators—9 kHz, 10 kHz and 11 kHz—would produce the same results as modulating a 10 kHz carrier with a 1 kHz sine wave?

6. Why is over-modulation normally considered to be bad?

7. Why is it desirable to keep the percentage modulation as high as possible?

8. For single sine wave 100% modulation, how much of the total signal power is contained in each sideband?

9. If a complex audio signal is used to amplitude modulate a carrier to 100%, does the power relationship calculated in question 8 still apply?

10. The output of an AM transmitter is a 3.5 watt carrier before modulation is applied. If 60% sine wave amplitude modulation is added:

   a. What is the carrier power?

   b. What is the power in each sideband?

   c. What is the total RF power output?

11. A class-C modulation/amplifier draws 1.2 amps from a 24 volt power supply when only the RF carrier is turned on. An audio amplifier adds 10 watts of modulating signal to the total power supply. The RF transistor used for the modulator is 58% efficient. What total RF power appears at the output, how much power is wasted as heat, and what is the percent modulation if the modulating signal is just a sine wave?

6. FREQUENCY AND PHASE MODULATION

The previous chapter showed how information could be carried as high-frequency signals by modulating the amplitude of a carrier. This chapter continues with modulation techniques by showing how frequency and phase changes of the carrier could also be used to carry information. Although frequency modulation (FM) and phase modulation (PM) are two different techniques, they have much in common and are best described in one chapter. Collectively, FM and PM are referred to as *angle modulation*.

This chapter will:

1. Examine the strength and position of the sidebands created by sine wave, square wave, and other modulating signals.
2. Briefly explore the system used to broadcast and receive compatible FM mono/stereo music.
3. Calculate bandwidth needed for various FM waves.
4. Show how phase modulation differs from frequency modulation.
5. Explain the effects of frequency multiplication on an FM signal.
6. Look at circuitry for both FM and PM modulators and demodulators.
7. Show how well FM and PM perform when they become noisy and whether they are really superior to AM.

Some of the important points about modulation of any kind are worth repeating here:

- When any characteristic of a sine wave is changed, new side frequencies appear. These will be symmetrically located above and below the carrier and spaced away from the carrier by an amount equal to the modulating frequency.
- The resulting sidebands collectively contain the same information as the original modulating signal.

- Multiple sidebands may be created and these provide information redundancy that gives some protection against errors caused by unwanted stray signals and distortion.

These points are equally true for FM and PM. In fact, their big benefit
over AM is that PM and FM can create large numbers of sidebands on each side
of the carrier for each modulating frequency. As a result, better performance in
the presence of small amounts of noise is obtained. However, angle modulation
(FM and PM) also has its faults and so all three modulation techniques exist. No
single technique is clearly superior to the others; they each have their areas of
usefulness.

Frequency modulation is used for FM stereo broadcasting (88 to 108 MHz)
where its noise suppression characteristics complement the music that is
broadcast. FM is also used for mobile police, fire, and taxi radios as well as
mobile radio telephones. The television sound channel uses FM as does the
video portion of the television satellite broadcasts in the 3.7 to 4.2 GHz band.
Video recorders (VCRs) use FM to record the luminance (brightness) signal.

For both FM and PM, the amplitude of the modulated signal remains constant at all times. The frequency and the phase, however, are interrelated and
they both change for each type of modulation. To understand the relation between FM and PM look at Figure 6.1.

This figure shows a sine wave that starts at a frequency that has been
arbitrarily labeled 1000 Hz. After a while, the wave suddenly shifts to a new
frequency which has been labeled 1200 Hz. It stays at this new frequency for
five cycles and then abruptly returns to its original 1000 Hz. There have been no
changes in the amplitude of the sine wave and at no time did any instantaneous
changes in voltage occur. The illustration happens to show that the frequency
changes occurred right on the zero crossing points of the sine wave but this
isn't always the case nor does it matter. The modulated sine wave has made
frequency changes that can be represented by the rectangular wave in (c).

To use this process in reverse, if we had a waveform like that at (c) and it
was used to frequency modulate a 1000 Hz sine wave, the results would look
like that in (b).

But, more than just the frequency changes; the phase also changes. Since
phase must always be measured relative to something, the constant reference
sine wave at (a) is added. (It is not part of the transmitted signal.) This reference stays at the original 1000 Hz to show what the phase angle would have
been if the frequency hadn't changed. For the first few cycles, both waves (a)
and (b) are at the same frequency and stay in phase. Then, when the modulated
wave increases to 1200 Hz, its phase starts to slowly move ahead of what it
would have been if it stayed at 1000 Hz. As more cycles pass, the phase advances further and further. Finally, after five cycles at the higher frequency, the
signal returns to its lower frequency and there is no further phase change. For
the values used, the waveform at (b) will now be 300° ahead of (a) and will stay

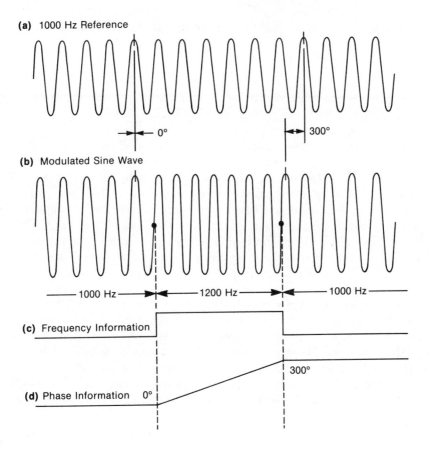

Figure 6.1 Frequency and phase changes occur at the same time for both modulation types.

at this constant value. The "ramp" waveform at (d) describes the phase changes that occurred along with the frequency changes.

If we had wanted to phase modulate a 1000 Hz sine wave using the ramp waveform at (d) then it too would look like the modulated sine wave at (b).

In summary, then:

- You cannot simply look at an angle-modulated waveform and determine whether it has been frequency or phase modulated.

- When a sine wave changes in frequency there will always be a corresponding change in phase.

- The frequency change and the phase change information are not the same shape. The PM information from the modulated wave is the mathematical integral of the FM information.

- For any given waveshape that is going to be used to angle modulate a sine wave there are two different circuits that can be used—a frequency modulator and a phase modulator. They will produce different results.

6.1 SIDEBANDS WITH SQUARE WAVE MODULATION

Now let us consider the frequency spectrum of the angle-modulated waveform to find out what sidebands are created. Initially, we will concentrate on frequency modulation. The easiest situation to start with is that shown in Figure 6.2. Here, a 1 kHz square wave frequency modulates a 100 kHz carrier. The frequency modulator inside the block could consist of a single oscillator whose frequency depends on the input voltage—in other words, a voltage-controlled oscillator (VCO) as described in Chapter 4. The signal source and the modulator are, therefore, combined in the same circuit. The square wave input makes the oscillator shift between a lower frequency of 90 kHz and a higher one of 110 kHz.

Our knowledge of AM waves and sidebands can now be used to help with this simple example of FM. The carrier has been shifted so that it spends equal time at 90 kHz and at 110 kHz. If two bandpass filters were to be used to examine the wave, the frequencies near 90 kHz could be separated from those near 110 kHz. The setup and its results are shown in Figure 6.3.

The output from each filter now appears as an amplitude-modulated wave. When the carrier is shifted to the lower frequency, an output appears

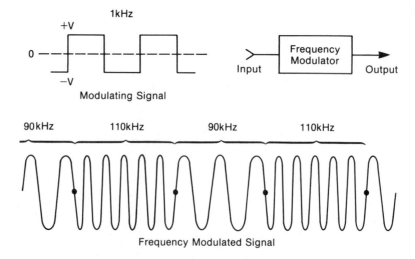

Figure 6.2 A square wave shifts the frequency of a 100 kHz oscillator. The amplitude of the output signal stays constant even though the frequency changes.

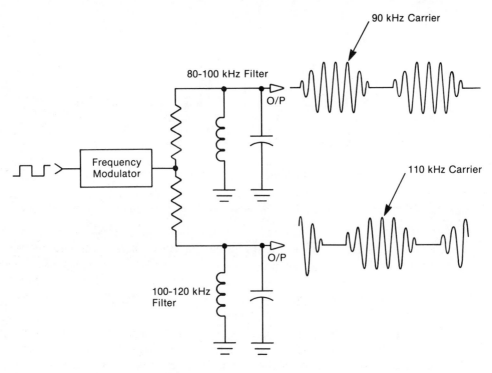

Figure 6.3 To analyze a simple FM wave, two bandpass filters can be used.

from that filter only. When the carrier is shifted to the higher frequency, the output from the lower-frequency filter slowly drops to zero and the output from the higher-frequency one rises to its maximum. The slow rise and fall are the results of the limited bandwidths of each filter. Having found the AM equivalence of this simple example, the frequency spectrum of the FM output can be drawn.

A square wave consists of a fundamental and all of the odd harmonics of this fundamental. For our 1 kHz square wave, frequency components with decreasing amplitude exist at 1, 3, 5, 7, 9, 11 kHz ... etc. The FM spectrum will, therefore, show two carriers with these sidebands above and below each carrier frequency. The result is shown in Figure 6.4.

Although the carrier existed at only two frequencies (90 and 110 kHz), the movement of the carrier has created sidebands above and below each position. The spectrum components are strongest near the two carrier positions and drop off in strength the further they move away from these two positions. With reference to this frequency spectrum, two terms can be explained:

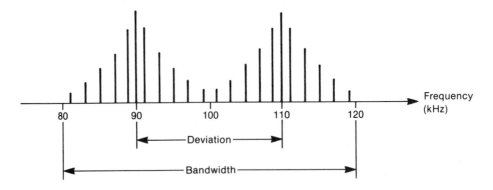

Figure 6.4 Frequency spectrum of an FM wave resulting from square wave modulation.

Deviation describes the frequency range (in Hz or kHz) over which the carrier is moved. For this example, the peak deviation is ± 10 kHz and the peak-to-peak deviation is 20 kHz. Almost any value of deviation can be used and we will see shortly the advantages and disadvantages of selecting different values.

Bandwidth is the total amount of frequency space (Hz or kHz) used up by all the sidebands. This value is always greater than the deviation but there is no fixed relation between them. The bandwidth for our example is 38 kHz. However, weak sidebands could theoretically exist over an infinite bandwidth.

6.2 SIDEBANDS WITH SINE WAVE MODULATION

The previous example was very simple in that the carrier, with square wave modulation, existed at only two discrete frequencies. A deviation of 10 kHz was also chosen to be much larger than the modulating frequency and this helped make the example simpler. A more general look at the sidebands of an FM signal is obtained when sine wave modulation is used. Now the carrier makes no instantaneous frequency shifts. Instead, it smoothly moves back and forth between its peak deviation frequency limits.

With a single sine wave modulating its frequency, the voltage at any instant in an FM wave is:

$$e = E_c \sin\left[\omega_c t + m_f \sin(\omega_m t)\right] \tag{6.1}$$

where E_c is the peak carrier voltage
  ω_c is the carrier's center frequency (radians/sec)

ω_m is the single modulating frequency (rad/sec)

m_f is the FM "modulation index"

$$m_f = \frac{\Delta f}{f_m} = \frac{\text{Peak carrier deviation (Hz or Rad/sec)}}{\text{Modulating frequency (same units)}} \qquad (6.2)$$

The *modulation index* is the FM equivalent of the "percent modulation" (*m*) used with AM.

The mathematical expansion of this formula is extremely difficult, mainly because the sine of another sine must be derived. The solution involves the use of Bessel functions. Without providing any intermediate steps, the final solution is shown here:

$$\begin{aligned}
e = E_c \Big\{ &J_0 \sin \omega_c t \\
&+ J_1 \left[\sin (\omega_c + \omega_m)t - \sin (\omega_c - \omega_m)t \right] \\
&+ J_2 \left[\sin (\omega_c + 2\,\omega_m)t + \sin (\omega_c - 2\omega_m)t \right] \\
&+ J_3 \left[\sin (\omega_c + 3\,\omega_m)t - \sin (\omega_c - 3\omega_m)t \right] \\
&+ J_4 \left[\sin (\omega_c + 4\,\omega_m)t + \sin (\omega_c - 4\,\omega_m)t \right] \\
&+ \ldots\ldots\ldots\ldots \Big\}
\end{aligned} \qquad (6.3)$$

This expansion shows that:

- Even with a single sine wave frequency modulating a carrier, an infinite number of upper and lower sidebands are formed (unlike AM).
- The sidebands are spaced symmetrically above and below the original carrier frequency by multiples of the modulating frequency (ω_m, $2\omega_m$, $3\omega_m$, etc).
- The amplitude of each pair of sidebands depends on its Bessel coeffieients—these are the J_0, J_1, J_2, etc., terms. Their values depend on the modulation index (m_f).
- The carrier itself changes amplitude because the J_0 coefficient is associated with it.
- The phase of the odd sidebands of the lower side of the carrier is the opposite of those on the higher side. The even sidebands do not show this difference.

The Bessel coefficients are numbers that change within the range 0.0 up to $\pm$ 1.0. They indicate the strength of each component relative to the amplitude of the original, unmodulated carrier. If a coefficient is zero then the corresponding pair of the sidebands will not exist. If a coefficient is negative then both sidebands in the pair will have the opposite phase (180° shift) that they normally would have.

The coefficient values are very dependent on the modulation index. Therefore, the modulating frequency and the deviation both affect the sideband strengths. This is different from AM where modulation depth was the only factor that determined the strength of the sidebands. Tables of coefficient values are available if great accuracy is desired; however, for our use their values can be read off the graphs of Figure 6.5. One advantage of the graphs is that they indicate the general trend of each coefficient as the modulation index changes.

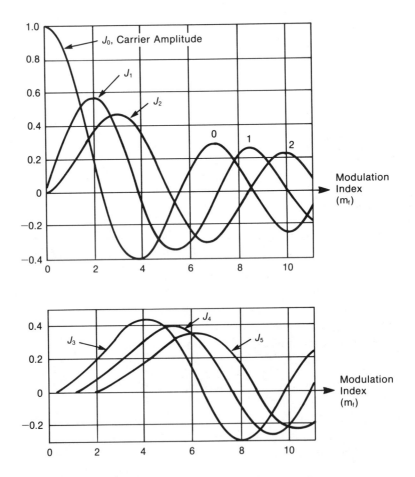

Figure 6.5 Bessel coefficient values for J_0 (the carrier) up to the fifth sideband (J_5). The coefficient values depend on the modulation index (m_f).

For some values of m_f some coefficients will be zero, some will be positive, and others will be negative. For small values of m_f only one or two sidebands will be significant; for larger values, a great number of sidebands will appear.

The following example will demonstrate the use of Bessel coefficients to determine the sideband content of an FM signal.

Example 6.1 ══

Determine the sideband components for an FM wave that results from a 2 kHz sine wave modulating the frequency of a 1.0 MHz sine wave. The amplitude of this carrier before it is modulated is 10 volts rms. The modulating signal is strong enough that the carrier is shifted between deviation peaks of $\pm$ 1.6 kHz.

Solution:

For the values given, the modulation index is:

$$= \frac{\text{peak deviation}}{\text{modulating frequency}} = \frac{1.6 \text{ kHz}}{2.0 \text{ kHz}} = 0.8$$

Using this value, the Bessel charts give the amplitudes of the significant sidebands.

$$J_0 = 0.846 \text{ (carrier)}$$
$$J_1 = 0.369 \text{ (sidebands at } \pm \text{ 2 kHz)}$$
$$J_2 = 0.076 \text{ (sidebands at } \pm \text{ 4 kHz)}$$
$$J_3 = 0.010 \text{ (sidebands at } \pm \text{ 6 kHz)}$$

These four values are the only ones higher than 0.01. While other sidebands do exist, their amplitudes are so small that they can be ignored without upsetting the resulting FM wave.

The resulting frequency spectrum is shown in Figure 6.6. It shows that the carrier has been reduced from 10 volts down to 8.46 volts. The missing energy now appears in the sidebands. The first pair of sidebands are 2 kHz above and below the carrier and have amplitudes of 3.69 volts. The lower sideband of this pair is shown with a negative value which indicates that it starts out with a 180° phase inversion relative to the carrier signal. The next pair of sidebands are 4 kHz above and below the carrier. They each have an amplitude of 0.76 volts and both start out in phase with the carrier. The last significant sidebands lie 6 kHz above and below the carrier and are very weak. Their amplitude is only 0.1 volt. Again, the lower sideband starts with an inverted phase and so is labeled with a negative value.

The modulating signal that causes these sidebands is included for comparison. It is shown as a 2 kHz sine wave that causes the carrier to deviate only 1.6

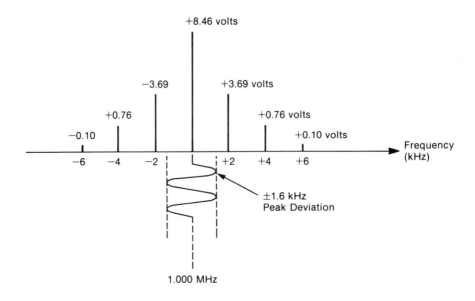

Figure 6.6 Frequency spectrum of an FM wave where a carrier is deviated to ± 1.6 kHz peaks by a 2 kHz modulating signal.

kHz above and below its original position. The resulting signal occupies a bandwidth of 12 kHz.

Figure 6.7 further demonstrates the amplitude and phase relation of the seven individual frequencies that make up the FM wave. The previous illustration (Figure 6.6) is similar to that seen on a spectrum analyzer. Phase relationships cannot be detected with a spectrum analyzer. However, if the analyzer's sweeping is stopped and each sideband is watched to see how it changes with time, the results of Figure 6.7 could be obtained. Now phase relationships can be shown.

At the top of the illustration, the 2 kHz modulating signal is shown for comparison. As it passes through its zero point, the modulated carrier will just be crossing through its center frequency of 1.000 MHz. All seven component signals are also shown just crossing through their zero points. Five of them are heading toward positive values initially while two of them are heading in the negative direction. These are the two that were marked with negative amplitudes in the previous illustration. If all seven sine waves are carefully added together point by point (a very tedious task), the result will be the frequency-modulated carrier with a constant amplitude of 10.0 volts. A simple addition of each component's amplitude will *not* produce a total of 10 volts, however. The reason is that all seven components are at different frequencies and two differ-

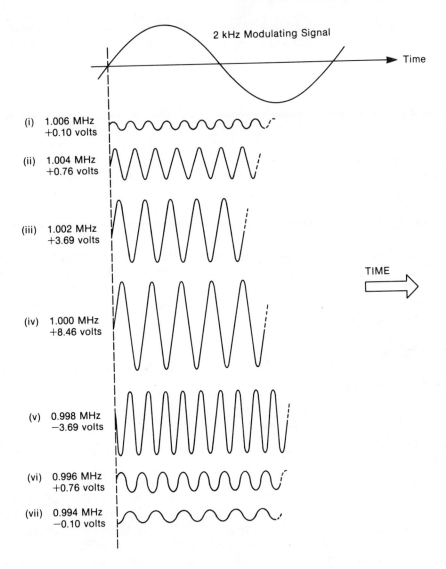

Figure 6.7 Seven individual sine waves make up the FM wave from Example 6.1. Notice that two of the sine waves start in the negative direction.

ent phases are involved and so direct addition is not possible. Addition that considers the + and − signs does work.

$$0.10 + 0.76 + 3.69 + 8.46 - 3.69 + 0.76 - 0.10$$
$$= 9.98 \text{ volts}$$

The correct total should be 10.0 volts. The small difference is caused by two things. Part of the problem is the small number of significant figures used in each of the Bessel coefficients. The other reason is that, as mentioned, there are more sidebands but these were ignored because their amplitudes were so low.

The next example uses a higher deviation with the result that one of the J terms becomes zero and the others have positive and negative values.

Example 6.2

For data communications, the transmitter portion of a modem has a carrier frequency of 1500 Hz. It is frequency modulated by a 100 Hz sine wave causing the carrier to reach a peak deviation of $\pm$ 550 Hz. The unmodulated carrier amplitude is 5.0 volts. Determine the amplitude and relative phase of each component of the frequency spectrum.

Solution:

The modulation index is:

$$\frac{\text{peak deviation}}{\text{modulating freq.}} = \frac{550 \text{ Hz}}{100 \text{ Hz}} = 5.5$$

This value is used to find the Bessel coefficients for each of the significant components.

Next, the amplitudes and relative phase angles of each frequency component are calculated. Remember that all of the upper sidebands *normally* start at zero phase angle as do the even ones on the lower side. The odd, lower sidebands *normally* start at a relative angle of 180°. The Bessel coefficients that have negative values will cause the phases of the corresponding sidebands to change by an additional 180°. In the following chart, which summarizes all the results, the (+) sign is used for zero phase angle and the (−) is used for the 180° starting phase.

Frequency	Bessel Coefficient	Normal Phase	Final Amplitude and Phase
600 Hz	$J_9 = +0.02$	−	−0.10 volts
700	$J_8 = +0.04$	+	+0.20
800	$J_7 = +0.10$	−	−0.50
900	$J_6 = +0.18$	+	+0.90
1000	$J_5 = +0.32$	−	−1.60

1100	$J_4 = +0.39$	+	+1.95
1200	$J_3 = +0.27$	−	−1.35
1300	$J_2 = -0.14$	+	−0.70
1400	$J_1 = -0.34$	−	+1.70
1500	$J_0 = 0.00$	+	no carrier component
1600	$J_1 = -0.34$	+	−1.70
1700	$J_2 = -0.14$	+	−0.70
1800	$J_3 = +0.27$	+	+1.35
1900	$J_4 = +0.39$	+	+1.95
2000	$J_5 = +0.32$	+	+1.60
2100	$J_6 = +0.18$	+	+0.90
2200	$J_7 = +0.10$	+	+0.50
2300	$J_8 = +0.04$	+	+0.20
2400	$J_9 = +0.02$	+	+0.10

The corresponding spectrum diagram is shown in Figure 6.8. The phases are indicated by the + and − signs on the component amplitudes.

Some further insight into FM and its sidebands is provided in the next two illustrations. Figure 6.9 shows four different frequency spectrums. Each is the result of modulating a carrier with a 2 kHz sine wave but with different amplitudes so that the peak carrier deviation changes. The result is that all sidebands are spaced apart by a fixed 2 kHz value but there is a great difference in the number of sidebands produced. Very small deviations result in only one significant sideband on each side of the carrier and this has the advantage of reduced bandwidth. FM communications systems that use this feature are called *narrow band FM* (NBFM) systems. The disadvantage of the small deviation is that very few sidebands are created and this limits the signal-to-noise ratio improvements that would otherwise be possible at the receiving end. NBFM is used for

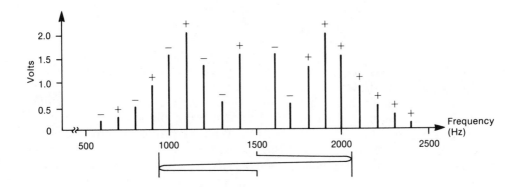

Figure 6.8 Sidebands for a 1500 Hz carrier deviated ± 550 Hz by a 100 Hz sine wave.

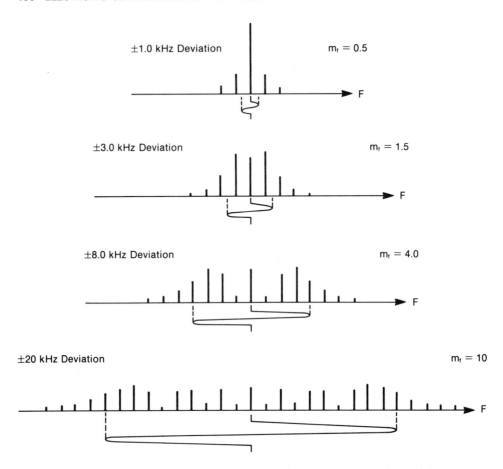

Figure 6.9 FM sidebands resulting from a constant modulating frequency of 2 kHz but with different peak deviations.

radio systems where communication is more important than noise-free music. Typical applications are police, taxi, and mobile telephone transceivers (±5 kHz peak deviation).

The same figure shows that wider deviations create an abundance of sidebands. These use up a great deal of bandwidth but the result is greatly improved signal-to-noise ratio at the receiver. Wider deviation systems are called *wideband FM* (WBFM) systems. WBFM is used for entertainment applications such as FM stereo broadcasting (±75 kHz peak deviation) and the sound channel of commercial television (±25kHz peak deviation) where the extra quieting is appreciated by the listeners.

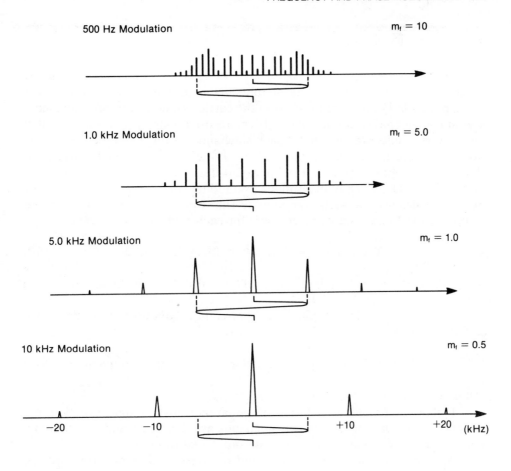

Figure 6.10   FM sidebands resulting from a constant peak deviation of ±5 kHz but with different modulating frequencies.

Figure 6.10 continues our description of FM and its sidebands. In this illustration, the peak deviation has been held constant at ±5 kHz and the modulating frequency is being changed. This situation could result if a variable frequency audio oscillator were connected to the input of an FM transmitter and used to measure its "frequency response." Notice that the bandwidth changes less than before but the number of sidebands changes drastically. A very important point to be remembered from this is that the higher frequencies do not generate as many sidebands and this, in turn, results in a poorer signal-to-noise ratio for the higher frequencies at the receiver. This problem will be dealt with a little later in this chapter.

6.3 FM BANDWIDTH

How much bandwidth is needed for an FM signal? Theoretically, an infinite bandwidth is needed. But many of the sidebands that are created are very weak, particularly those far away from the carrier. Good results are obtained if a group of sidebands around the carrier are used. One criteria suggests that satisfactory results are obtained if each sideband, starting from the carrier and working outward, is included until at least 98% of the total signal power is included. All sidebands past that point would not add any significant improvement. The total power is determined by adding together the square of the Bessel coefficients for each component. This would include the one carrier component (J_0) and then two sidebands for each remaining coefficient.

$$\text{Total power} = J_0^2 + 2J_1^2 + 2J_2^2 + 2J_3^2 + 2J_4^2 \tag{6.4}$$

If all sidebands are included, this total will be 1.000; but, satisfactory operation is achieved with a total of 0.98.

The 98% power limit results in an approximate bandwidth formula known as Carson's Rule:

$$\text{FM bandwidth} \approx 2 \text{ (peak deviation + modulating frequency)} \tag{6.5}$$

(All parts of this equation are measured in Hz.)

It should be remembered that this equation is just an approximation that suits one particular guess at which sidebands are significant and which aren't. Wider or narrower bandwidths may suit individual situations.

Another important point that should be considered along with a discussion of bandwidth is the delay characteristics of any filters that are used to set this bandwidth. (Delay characteristics were discussed in Chapter 3.) Since each modulating frequency could produce a large number of sidebands, the phase relation between these sidebands is critical. Any filters and amplifiers used must have a reasonably flat delay response across the bandwidth occupied by the FM spectrum. FM is not as tolerant of delay distortion as AM because AM creates only one pair of side frequencies for each modulating frequency.

Example 6.3 ━━━

What bandwidth is needed for an FM signal that has a peak deviation of ± 3 kHz and handles audio signals from 200 to 5000 Hz?

Solution:

Using Carson's Rule for the lowest and then the highest modulating frequency:

$$BW_1 \approx 2\ (3 + 0.2)\ \text{kHz}$$
$$= 6.4\ \text{kHz}$$
$$BW_2 \approx 2\ (3 + 5)\ \text{kHz}$$
$$= 16\ \text{kHz}$$

This second value is the wider of the two bandwidths so any filters for this signal would need a minimum bandwidth of 16 kHz.

6.4 COMPLEX MODULATING SIGNALS

So far, the description of FM sidebands has centered around modulation by only a single sine wave. What happens if a more complex wave is used where many different modulating frequencies are present at the same time? The result is rather complex. A similar problem with AM was very easy—the total sideband pattern was simply the sum of all the sidebands created by each modulating frequency acting separately. With FM, this approach is not possible. FM is a "nonlinear" form of modulation that involves intermodulation products similar to those discussed with nonlinear amplifiers back in Chapter 1. There will be sidebands at each of the frequencies suggested by the one-at-a-time approach but there will be a large number of additional sidebands appearing at the sum and difference frequencies of all the modulating frequencies and their harmonics. Amplitude calculations for each sideband are complex since they involve not only Bessel coefficients but also the amplitude of each modulating frequency involved. In short, the result is a mess that is not easily calculated.

The sidebands show a "tendency" to be strongest at frequencies where the carrier signal is shifting the slowest. This was illustrated in Figure 6.1 and is further described in the Figure 6.11.

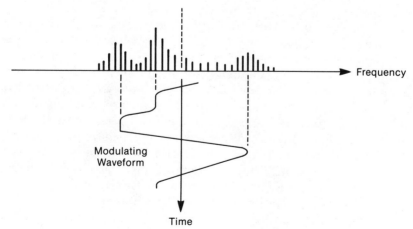

Figure 6.11 When a complex modulating signal is involved, sidebands tend to be strongest where the carrier shifts the slowest.

6.5 FM STEREO BROADCASTING

An excellent example of FM in everyday use is the commercial broadcasting service in the 88 to 108 MHz band. This example has an added benefit because those stations that transmit a stereo signal include an extra AM signal on a subcarrier and this will also be described.

The FM band has about 20 times more frequency space than the 540 to 1600 kHz AM band even though each band has about the same number of available channels (about 100). Therefore, each FM transmitter can afford to use up a wider bandwidth so wide-deviation FM is used (WBFM). The channels are assigned every 200 kHz starting at 88.1 MHz. This tends to make it look as if each station can use only its allocated 200 kHz. However, as we shall see, more can be used without causing serious interference to stations in the adjacent channels.

The audio signals carried by each station are limited to the frequency range 50 Hz to 15,000 Hz. Although this may seem less than true Hi-Fi, it is perfectly adequate for all but the most discerning listener. No single audio frequency drives the carrier to its full deviation. Instead, the transmitter is adjusted so that only the peaks of the complex audio waveform cause the full $\pm$ 75 kHz deviation that is legally allowed. Using Carson's Rule, the approximate bandwidth for this signal will be:

$$
\begin{aligned}
BW &\approx 2 \ (75 \ \text{kHz} \ + \ 15 \ \text{kHz}) \\
&= 2 \ (90) \\
&= 180 \ \text{kHz}
\end{aligned}
$$

This value fits nicely within the 200 kHz assigned channels.

The system as described so far will work but there is a problem. Each audio frequency could cause equal amounts of carrier deviation. If this were allowed, the lower modulating frequencies would produce large numbers of sidebands (large value of m_f) and the higher frequencies would produce fewer sidebands (low value of m_f). Since the resulting signal-to-noise ratio at the receiver is proportional to the number of sidebands, the higher frequencies will have a tendency to be noisy. Compounding the problem is the fact that the average music signal has relatively weak (but still important) high frequencies and these audio frequencies cause less than their share of carrier deviation. This problem was already mentioned when Figure 6.1 was discussed. The solution is to make the lower frequencies weaker and the high frequencies stronger before they are transmitted. This is called *high-frequency pre-emphasis* or simply *pre-emphasis*. After receiving, of course, the signal levels must be rebalanced again or the music would appear to be lacking in bass response. A corresponding high-frequency "de-emphasis," therefore, takes place in the receiver. This means that high frequencies are attenuated and this includes the high-frequency noise. The overall result is a flat frequency response as far as the audio is concerned and a lower amount of noise. A block diagram of the system is shown in Figure 6.12.

The pre-emphasis starts at 2122 Hz and increases the higher-frequency amplitudes at the rate of 6 dB per octave. This is shown in Figure 6.13. At 15 kHz, the highest audio frequency broadcast, the boost has reached + 17 dB. The network consists of the parallel R_1C_1 combination that sets a constant loss at lower frequencies. The "corner frequency" or 3 dB point for such a circuit is given by:

$$F_c = \frac{1}{2\pi R_1 C_1} \text{ (Hz)} \qquad (6.6)$$

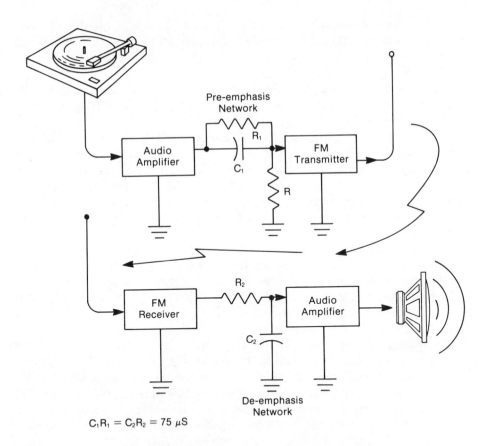

$C_1R_1 = C_2R_2 = 75 \ \mu S$

Figure 6.12 Pre-emphasis is used at the transmitter to improve high-frequency signal-to-noise ratio. A corresponding de-emphasis is needed at the receiver.

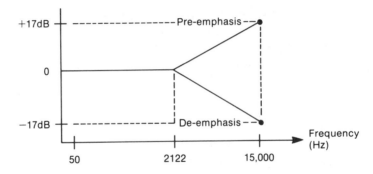

Figure 6.13 Frequency response of the 75 μ sec pre-emphasis and de-emphasis networks.

For a corner frequency of 2122 Hz, there are an infinite number of resistors and capacitors that could be used but the product of each RC combination will always be a "time constant" of 75 μsec. As a result, the network is often called a 75 μsec pre-emphasis circuit.

The de-emphasis network interchanges the resistor and capacitor positions to provide a low-pass characteristic. The 2122 Hz corner frequency and the 75 μsec time constant remain unchanged.

When stereo was introduced in 1961, many monophonic receivers already existed. It was essential that the new stereo service remain compatible with the older monophonic service. This meant that, although music was to be broadcast in stereo, owners of the older receivers must still receive the full music, not just the left channel or the right channel but both combined. There should also be very little loss in monophonic noise immunity. The system shown in Figure 6.14 resulted.

Sound is picked up by two microphones that provide one signal from the left side of the musical group and another from the right. After amplification and the normal pre-emphasis, a portion of each signal is added together (two equal-value resistors). This forms a LEFT-plus-RIGHT (L + R) or "sum" channel that provides the required monophonic compatibility. Then one channel is inverted and this signal is added to the uninverted signal from the other channel. This forms a LEFT-minus-RIGHT (L − R) or "difference" channel. Only people with stereo receivers will need this signal. The difference channel is carried at a higher modulating frequency so that it doesn't interfere with the (L + R) portion. Double sideband suppressed carrier modulation is used with a 38 kHz carrier. To aid in synchronous demodulation at the receiver, a weak 19 kHz "pilot carrier" is also included in the transmission. The receiver uses the second harmonic of the pilot to demodulate the difference channel.

Why not just include a weak 38 kHz carrier? By not including the proper carrier the possibility of accidental AM demodulation taking place in older sets is reduced. This would be possible if any amplifier had a slight nonlinearity and

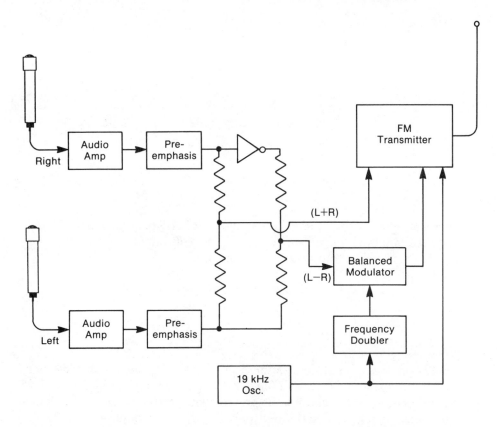

Figure 6.14 Stereo encoding at the FM transmitter creates a sum channel that travels as audio and a difference channel that travels on a suppressed 38 kHz subcarrier.

the demodulated signal would then make one channel weaker and the other stronger.

The frequency spectrum of the composite signals before they go to the FM modulator is shown in Figure 6.15. A sample composite waveform is also shown for the situation where the left channel is a triangular wave and the right channel is a sine wave.

At the receiver, after FM demodulation takes place, the same frequency spectrum and waveshape reappear. If monophonic listening is desired, the higher-frequency signals are ignored and only the (L + R) "sum" channel is used. If stereo operation is to occur, the difference channel is needed. First, the 19 kHz pilot tone is picked out of the composite signal using a narrow bandpass filter. Next, it is passed through a nonlinear amplifier that creates harmonics and the 38 kHz second harmonic is picked out with another bandpass filter. Finally, this regenerated carrier is used to synchronously demodulate the DSBSC difference channel. Two audio frequency signals now exist; one is the

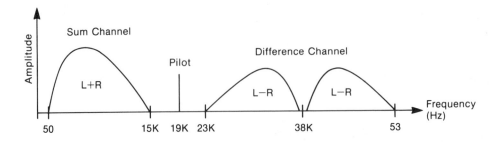

Figure 6.15 Composite stereo frequency spectrum.

sum and the other is the difference channel. The original LEFT and RIGHT channels are obtained by the same adding and subtracting (invert and add) process that was carried out at the transmitter.

$$\text{LEFT} = \text{sum channel} + \text{difference channel} \qquad (6.7)$$
$$(L + R) \qquad + (L - R)$$
$$\text{RIGHT} = \text{sum channel} - \text{difference channel} \qquad (6.8)$$
$$(L + R) \qquad - (L - R)$$

How well does the FM stereo system work? As far as monophonic listeners are concerned, their signal became only 1 dB weaker than it was before. This was mainly caused by the continuous presence of the 19 kHz pilot. The signal-to-noise ratio, therefore, deteriorated only 1 dB for monophonic operation. Stereo operation is considerably noisier. Because the difference channel at the transmitter is held at a weaker level than the sum channel and because it is

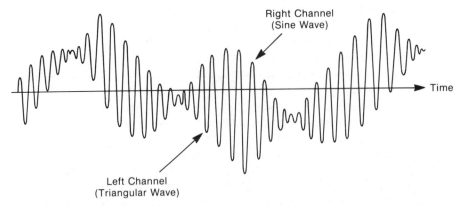

Figure 6.16 Waveshape of the composite stereo signal with one channel carrying a triangular waveform and the other a sine wave.

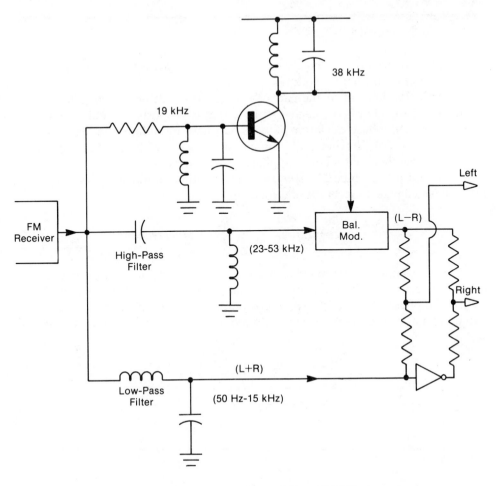

Figure 6.17 After FM demodulation the original LEFT and RIGHT channels are reconstructed.

carried at a very high modulating frequency (23 to 53 kHz), very few FM side-bands are created. The difference channel is, therefore, noisier than the sum channel. The resulting stereo signal that comes out of the speakers is about 21 dB noisier than the monophonic signal from the same transmitter. Most receivers, as a result, have a mono/stereo switch so that listeners can still enjoy music (in mono) from weak stations when the noise level is too high for good stereo operation.

6.6 PHASE MODULATION

Now it is time to look at phase modulation to see how it works, what its sidebands are like, and how it compares to FM.

With PM, instead of the carrier's frequency shifting proportional to the modulating signal, it is the carrier's phase that is proportional. It is possible to define the peak phase deviation values and phase modulation index similar to those used for FM. However, it is much easier to work with the FM terms themselves. Remember that when any phase change occurs, an accompanying frequency shift also results. The frequency shift is the mathematical derivative of the phase change as given by the following:

$$\text{Frequency (Hz)} = \frac{1}{360} \times \frac{d\Theta}{dt} \text{ (degrees/second)} \qquad (6.9)$$

Two examples are provided in Figure 6.18 that demonstrate the relationship between frequency and phase deviation. In example (a), a square wave shifts the phase of a carrier. The corresponding frequency shift is a differentiated version of the square wave—only momentary frequency shifts occur and

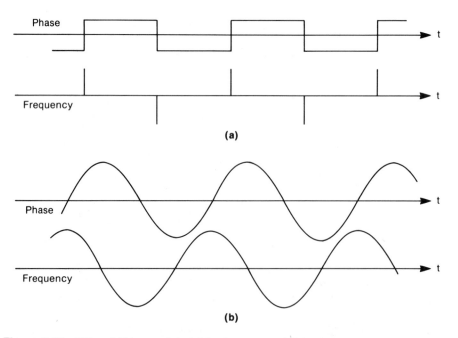

Figure 6.18 PM and FM are related. The frequency shift is always the mathematical derivative of the phase shift.

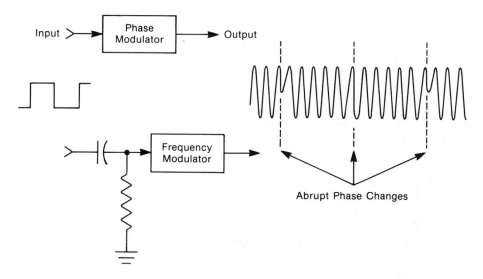

Figure 6.19   Phase modulation of a carrier can be obtained with either a proper phase modulator or with an FM modulator preceded by a differentiating network.

the carrier returns instantly to its middle value. The second example (b) shows a sine wave phase modulating the carrier. The corresponding frequency change is also a sine wave, but notice that it is shifted 90°—the result of the differentiating relationship.

There are two ways to phase modulate a carrier. One way is to use a proper phase modulator (to be described shortly). The other is to use a frequency modulator preceded by a differentiating circuit as shown in Figure 6.19. A differentiating network is a high-pass *RC* filter with its 3 dB "corner frequency" set higher than any frequency in the audio range. All audio frequencies will, therefore, lie on the + 6 dB per octave slope and experience a 90° phase shift. Higher frequencies will come out of the network stronger than the weak signals.

The differentiating network operates in a manner similar to the pre-emphasis network added to FM broadcasting. That network boosted only frequencies that were higher than 2122 Hz. The differentiating network must boost *all* frequencies at + 6 dB per octave. This means that the higher audio frequencies in the FM system were actually phase modulating the carrier.

For a constant amplitude signal into a phase modulator, the resulting frequency deviation will increase proportional to the modulating frequency. In other words, a low audio frequency will cause a small deviation and a higher frequency will cause a much greater deviation. The frequency modulation index (m_f) will be the same for all modulating frequencies so the same number of sidebands will be formed for each frequency. This is shown in Figure 6.20.

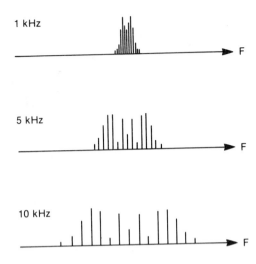

Figure 6.20   PM sidebands for three modulating frequencies. The input amplitude is constant so the phase deviation is also constant. The frequency deviation, however, increases and the number of sidebands remains the same.

Do you remember why pre-emphasis was added to the FM modulator? It was to get around the problem that resulted with "ordinary" FM where fewer sidebands were being created at higher modulating frequencies. This resulted in lower noise immunity for these frequencies. The advantage of PM over FM is, therefore, a constant S/N ratio for all modulating frequencies. However, this advantage is obtained at the expense of bandwidth. Higher frequencies with their higher-frequency deviations and large numbers of sidebands use up a lot of bandwidth.

Phase modulation is often used for higher-speed data modems where computer information must be sent over noisy telephone channels and error rates must be kept very low.

6.7   FREQUENCY MODULATORS

The obvious approach to frequency modulation generation is to use a voltage-controlled oscillator—perhaps similar to that shown in Figure 6.21, although many different circuits are available. The oscillator's characteristics must show a linear change of frequency with changes in applied voltage otherwise modulation distortion will result. Since frequency deviation, for most high-frequency applications, is small compared to the operating frequency, linearity isn't too difficult to achieve. For audio frequency modems with, for example, a $\pm$ 300 Hz

deviation and an 1800 Hz center frequency, linearity will be a little more difficult to achieve.

The simple VCO approach does present a problem—it will have a much
higher amount of center frequency drift than a crystal oscillator. This could be
a serious problem in VHF-FM operations where, for example, a transmitter
might have to operate at 160 MHz with a center frequency stability of 1 kHz
and a peak deviation of $\pm$ 5 kHz. There are several solutions to the drift problem. A common approach is to use two varactor diodes with the VCO. One is
fed with the audio-modulation voltage and the other is fed with a DC voltage
that changes very slowly to correct any frequency drift. This drift could be detected by comparing the VCO's average frequency to a reference crystal and
then creating a DC voltage that is proportional to the frequency difference. The
circuitry necessary to do this will be discussed along with phase-locked loops in
Chapter 9.

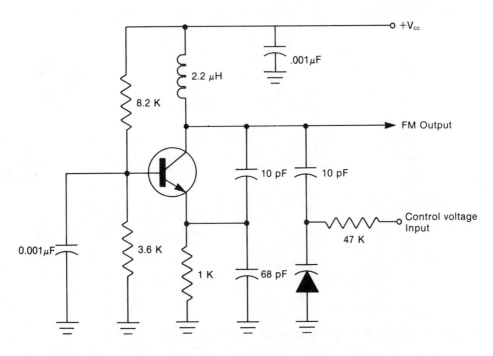

Figure 6.21   Voltage-controlled oscillator for high-frequency generation of FM.

6.8 PHASE MODULATORS

Phase modulator circuits are much easier to make frequency-stable since the modulating voltage doesn't affect the oscillator itself. The modulator circuitry is added after the oscillator in much the same way as amplitude modulation. The oscillator could, therefore, be high-stability crystal-controlled.

The phase modulator can take a number of forms but they will all form some type of filter circuit. A low-pass form is shown in Figure 6.22 along with its amplitude and phase response. For a constant filter response, a variable phase shift is obtained when different frequencies are used. In the same manner, if the frequency is held constant and the cutoff frequency of the filter is changed, a variable phase shift will also result. The varactor diode provides the variable filter characteristic. As long as the cutoff frequency remains higher than the oscillator's frequency there will be very little amplitude change. Phase shift values between about 10° and 100° would be possible with this circuit.

6.9 ARMSTRONG PHASE MODULATOR

A second method of phase modulation borrows a circuit from the previous chapter—a balanced modulator. This method produces a very good approxima-

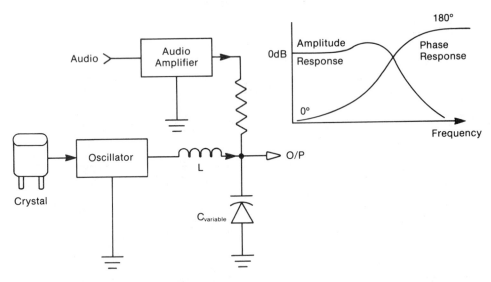

Figure 6.22 Phase modulator is added after a stable oscillator. It generally takes the form of a filter whose frequency and phase response is shifted with a varactor diode.

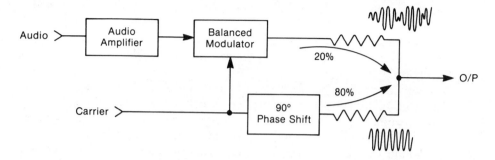

Figure 6.23 Armstrong modulator uses a balanced modulator to produce narrow-phase deviation PM.

tion of true phase modulation but only if the peak phase deviation is kept quite small.

The carrier is first used to produce a double sideband suppressed carrier signal and then the carrier is put back in. This may sound like a total waste of circuitry that might produce ordinary amplitude modulation. The significant difference is the 90° phase shift that takes place before the carrier is reinserted but after it is used to generate the sidebands. The phasor diagram of Figure 6.24 will explain the significance of this shift.

This figure represents modulation by a single audio frequency. A pair of sidebands is formed in the balanced modulator—one sideband will be higher than the carrier frequency and the other lower. These frequency differences cause the sideband vectors to rotate relative to the carrier's frequency vector. The two vectors produce a short resultant that is always at 90° to the reinserted carrier. When added to the carrier, the larger resultant is then a new vector that

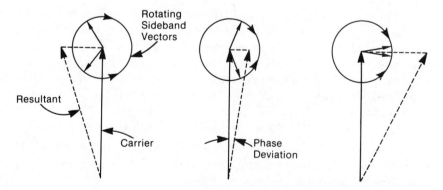

Figure 6.24 A DSB-SC signal added to a strong carrier that has been shifted 90° results in a good approximation of phase modulation.

has almost a constant amplitude and whose phase angle, relative to that of the original carrier's, changes almost directly proportional to the modulating voltage. If the 90° shift had not been used, the short resultant would have been in line with the carrier and would have caused an amplitude variation but no phase variation.

You will notice that the word *almost* appeared several times in the previous paragraph. If the DSB-SC signal is kept much smaller than the reinserted carrier, the results are very close to pure phase modulation. This restriction keeps the phase deviation to a small angle that is consistent since only a single pair of sidebands appears from this circuit for each modulating frequency. Wider phase deviation would require multiple pairs of sidebands and this circuit won't produce them.

The Armstrong method can also be used to produce narrow-frequency deviation FM. The same modulator is used but the audio signals are first passed through an integrator that reduces higher frequencies at the rate of −6 dB per octave. The result is FM with a stable average frequency since it is based on a crystal oscillator.

6.10 FREQUENCY MULTIPLIERS

A common circuit in many transmitters is a frequency multiplier. This simply consists of a nonlinear amplifier with a resonant circuit at its output that is tuned to a harmonic of the input signal. The amplifier is usually a single transistor operated without any base bias (class-C). Therefore, it will conduct only when the positive peaks of the input signal exceed 0.6 volts. Collector current will flow only in short pulses and the resulting distortion will contain many harmonics. Frequency multiplication by amounts of 2, 3, 4, and 5 are common.

What would happen if an FM or PM signal were fed into a frequency multiplier?

An FM or PM wave contains many frequencies at the same time. If such a complex signal is placed at the input of a multiplier, mixing between frequency components as well as harmonic generation will occur. The output signal will, therefore, contain more frequencies than the input signal even in the vicinity of the selected harmonic. This shouldn't be too surprising because this property of nonlinear amplifiers was described back in Chapters 1 and 5.

A specific example of an FM signal fed into a tripler is shown in Figure 6.26. The input has a 1.0 MHz center frequency and is deviated ± 1 kHz by a 2 kHz sine wave. This represents a modulation index of $m_f = 0.5$. A table of Bessel coefficients was consulted to obtain the frequency components for the input spectrum. Now for the output spectrum. When the input wave is at exactly 1.0 MHz, the tripled output would be at exactly 3.0 MHz. Then, at the instant the input has moved 1 kHz lower than 1.0 MHz, the output will be exactly 3 kHz

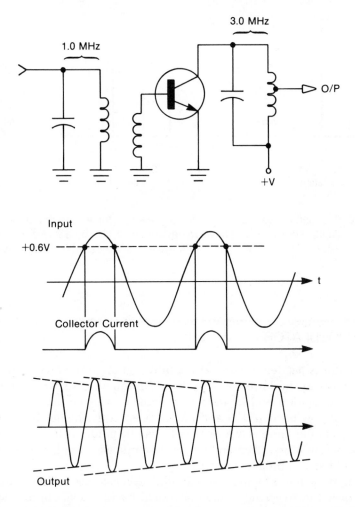

Figure 6.25 Class-C frequency tripler. Output energy decays slightly between input pulses.

below 3.0 MHz. Since the rate of deviation will still be the 2 kHz audio modulation, we see that the modulation index for the output wave will have increased. It will now be $m_f = 1.5$ because the deviation has tripled to ± 3 kHz. Again, the Bessel coefficient table was consulted to draw the output spectrum. Notice that many new signals have appeared and all the amplitude relations have changed.

An important application of frequency multipliers is to increase deviation values and modulation indexes for narrow-band FM and PM transmitters. This means, for example, that the Armstrong modulator which was limited to small deviations can still be used in wide-deviation transmitter circuits if frequency multipliers are added after the modulator.

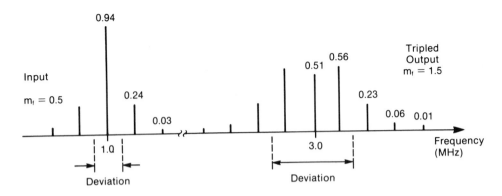

Figure 6.26 Frequency spectrum of the input wave to a tripler and the output wave. The input represents an FM wave deviated ± 1 kHz by a 2 kHz modulating signal. The output is still an FM wave but the deviation has increased to ± 3 kHz.

6.11 DEMODULATORS

A demodulator is needed to convert the information in the high-frequency carrier back down to the lower or "baseband" frequency. As we have seen with amplitude modulation and its detectors, some sort of nonlinear component—usually a diode—is needed. FM and PM, at least at the higher frequencies, are demodulated in the same way. Corrections can be made with the use of integrators and differentiators. (Remember the de-emphasis circuit needed at the receiver for FM-stereo broadcasting.)

One approach to FM and PM demodulation is to use a filter with a sloping response over the frequency range occupied by the deviating carrier and its sidebands. This will create an amplitude variation that should be roughly proportional to frequency and so a normal AM "envelope" detector can then be used.

This approach is quite workable but the problem with the simple-tuned circuit shown in Figure 6.27 is linearity. The response of the skirt of the filter does not fall linearly with frequency. Notice that the vertical and horizontal axes of the graph are both logarithmic. The demodulated signal would then be a distorted representation of the transmitted one.

A more linear filter is shown in Figure 6.28. The combination of a series and a parallel resonant frequency provides a straighter and also a steeper slope. The steepness helps because a small frequency variation now will result in a greater amplitude variation and so a stronger audio output will appear. For small frequency deviations at higher carrier frequencies, a quartz crystal or ceramic resonator can replace the tuned circuit.

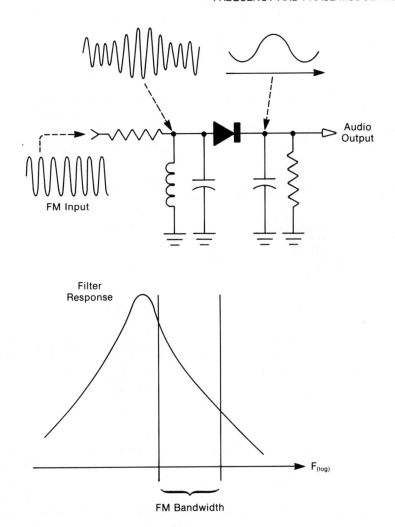

Figure 6.27 The simplest FM detector uses the sloping skirt of a tuned circuit to create an amplitude variation. An "envelope" detector is then used to perform the actual demodulation.

For wide deviation, as used in FM stereo broadcasting, the linearity still isn't adequate. Distortion of much less than 1% is needed. The balanced slope detector of Figure 6.29 uses two resonant circuits, each with its own AM detector. The output is the difference between the two detector voltages and, with careful design, can be made very linear. One circuit is resonant below the lowest sideband frequency and the other is resonant above the highest sideband. The resulting output is shown as the dashed line running between the peaks of the two curves. Its shape resembles the letter S.

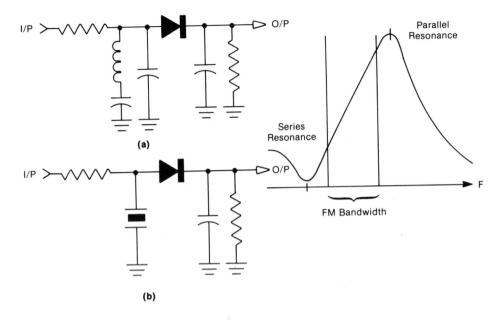

Figure 6.28 (a) An improvement in linearity and a stronger audio output are obtained with a series-parallel resonant circuit. (b) A quartz or ceramic resonator can also be used.

A further improvement is shown in Figure 6.30. Both the primary- and the secondary-tuned circuits are now resonant at the same frequency which is the center frequency of the FM (or PM) signal. This makes the initial circuit alignment at the time of manufacturing much easier. The same S-shaped response curve is now the result of two voltages added as vectors. One voltage (V_a) is capacitively coupled from the primary to the secondary and results in a voltage across the inductance (L). The other voltage is magnetically coupled into the secondary winding and, because of the center-tapped winding, results in the two voltages (V_b and V_c). Because the coupling between primary and secondary is intentionally kept low (small value of K), a 90° shift is created between the primary voltage and the secondary voltage. The shift will be slightly more or less than this, depending on the exact frequency, since the secondary is really a resonant circuit and so has a variable impedance. The result is summarized in the three small vector diagrams of the figure. At the center frequency, the primary to secondary shift is about 90° so, when added to the capacitively coupled voltage, two equal length vectors (V_d and V_e) are obtained. The two voltages are rectified by the diodes to create two equal DC voltages with opposite polarities. The resulting output is the sum of the two voltages and is zero.

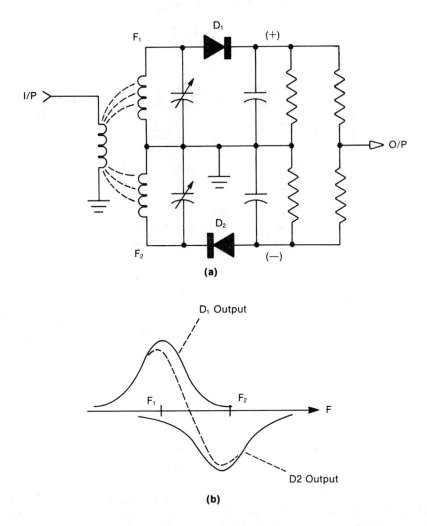

Figure 6.29 Balanced FM demodulator (a) and its frequency conversion curve (b).

At a frequency lower than the center frequency, the magnetically coupled voltage into the secondary will have a phase shift that is less than 90°. The center tap splits this again into two equal but opposite values and these are separately added to the capacitively coupled voltage. One of the resulting vectors is now longer than the other and, after rectification, two unequal DC voltages result. Depending on polarities, this could result in a net positive output.

In the same manner at a higher frequency, the phase shift will be greater than 90° and again will produce two vectors of unequal length. The resulting DC voltages could then create a final output that is negative. In summary, the

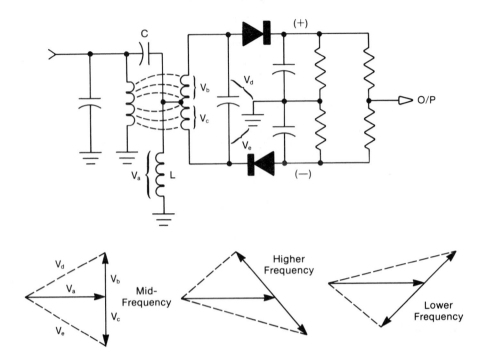

Figure 6.30 Foster-Seeley discriminator.

circuit operates by turning a frequency shift into a phase shift. When added to the original primary voltage, two new voltages of different amplitude appear. The diode rectifiers are not phase sensitive and produce a DC voltage proportional to the peak of each total voltage. Since one diode is reversed, the two voltages have opposite polarity and one voltage tends to cancel the other. The final output can then have either a positive or a negative value depending on which voltage is stronger. For constant amplitude of the input signal, the resulting DC output will be proportional to frequency deviation.

The Foster-Seeley discriminator and a companion circuit, the Ratio-Detector, have been used in FM tuners for many years.

All of the previous demodulator circuits have a slight deficiency; they are all sensitive to amplitude variations in the incoming FM or PM signal. This means that the demodulated audio output will be louder if the RF signal is stronger. It also means that if any noise is picked up along with the RF signal that noise will appear in the audio. To reduce the problem, the demodulator can be preceded by a limiter/amplifier. This could be a series of transistor amplifiers that are driven in and out of cutoff by signal or it could simply be a pair of parallel diodes with one reversed. The limiter is *not* an inherent requirement of all FM systems. It simply makes up for deficiencies in the demodulators.

Noise pulses that contaminate the RF signal will usually cause amplitude variations as well as phase variations. The limiter will, therefore, not be able to totally eliminate noise. The "low-noise" feature of FM and PM is provided by its
multiple sidebands created at the transmitter rather than by a set of limiters at
the receiver. A full FM receiver will be described in Chapter 8.

Another minor problem with some of these circuits is the special tuned
transformers needed. They make the circuitry bulky and more expensive. The
quadrature detector of Figure 6.31 is more suited to integrated circuit
fabrication and needs only a single inductor. This is a very popular circuit for
the sound channel of television receivers.

The input to this FM detector could be either a sine or square wave. A
preceding limiter is not absolutely necessary since this circuit is also self-limit-
ing as long as the input signal exceeds a certain minimum strength. The Exclusive-OR circuit for this application is made with pairs of differential amplifiers
as shown in Figure 6.32. Its logical function is the same as the normal gate used
with computers, etc.

In the detector circuit, the two capacitors C_1 and C_2 and the inductor L
form a network with a phase shift that changes rapidly with frequency. At the
center frequency of the FM signal, the phase shift will be about 90°. This is the

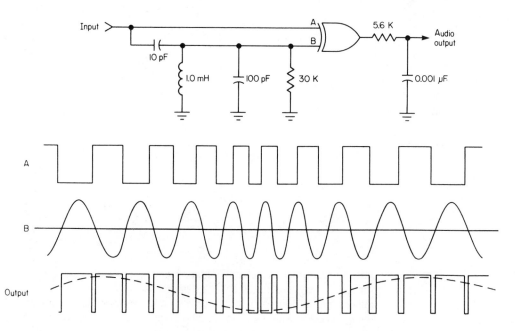

Figure 6.31 Quadrature FM detector uses an Exclusive-OR gate as a phase detector
and a network with a phase shift proportional to frequency. Component
values are for 455 kHz.

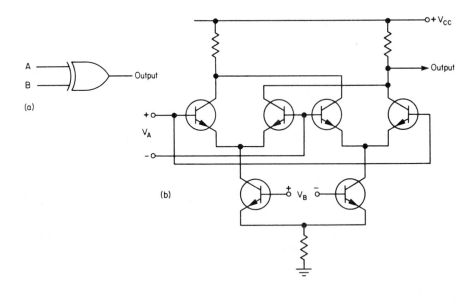

(a)

(b)

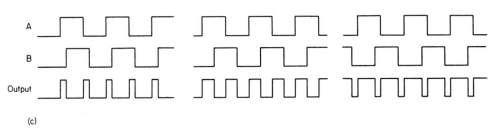

(c)

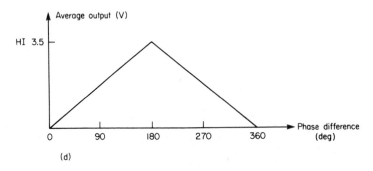

(d)

Figure 6.32 Phase detector using double differential amplifiers (a) has the same logic function as an Exclusive-OR gate (b). Its output consists of a series of pulses with a width proportional to input phase difference (c). The average output voltage change with input phase difference is summarized in (d).

small detail that gives the circuit its name—*quadrature*, at a 90° angle. For higher and lower frequencies, the shift will be correspondingly higher or lower. With one input to the logic gate being shifted and the other not, the output will be a series of pulses with a width proportional to the phase difference. The average value of these pulses will be proportional to the width since their amplitude isn't changing so we have a circuit that converts a small frequency shift into a DC voltage. The lower diagram in Figure 6.32 shows the relation between phase difference and output average voltage for the gate itself. With a 90° difference as the starting point, the input can shift an additional $\pm$ 90° about this point and still provide a linear, nonambiguous output.

The Exclusive-OR gate by itself is a good phase detector as long as the phase difference is limited to the 0° to 180° range. One of its inputs would be connected to a reference phase and its other input would be connected to the variable-phase signal. A very common application of the phase detector is in color television. The added signal used for the color information is contained on a 3.58 MHz subcarrier. The phase of this signal describes the color of the television screen at any instant and the amplitude of the subcarrier tells how strong the color will be. A black and white receiver simply ignores this subcarrier and uses only the "brightness" information that is also a part of the transmitted signal. The color receiver needs both the brightness and the color information. A modern receiver will often contain three of these detectors—one for the RED information, one for the GREEN, and one for the BLUE, these being the colors used by the phosphor dots on the screen of the picture tube. Each phase detector is fed with a reference signal that is derived from a reference "burst" sent out periodically by the transmitter. Figure 6.33 is a very simple representation of the three phase detectors decoding the color information.

6.12 FM AND NOISE

One of the advantages claimed for FM and PM is their superior noise immunity as compared to amplitude modulation. This immunity is due to the extra sidebands created during the modulation process but has the disadvantage of increased bandwidth. Just how much better is FM, for example, and are there any problems? The answer is provided by Figure 6.34. This compares the S/N ratio of a number of modulated signals just before they go into the appropriate demodulator with the resulting S/N ratio of the audio output.

The better modulation systems will have a higher S/N ratio after demodulation than they had before. This will be the result of some automatic correlation that takes place between the multiple sidebands during the demodulation process. The higher S/N ratio means a stronger audio signal and lower noise and so a much more pleasing signal to listen to or television picture to watch, etc.

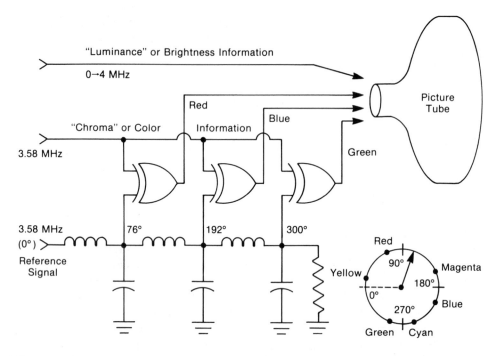

Figure 6.33 Three phase detectors are used to convert the 3.58 MHz subcarrier into the RED, GREEN, and BLUE signals needed for color television.

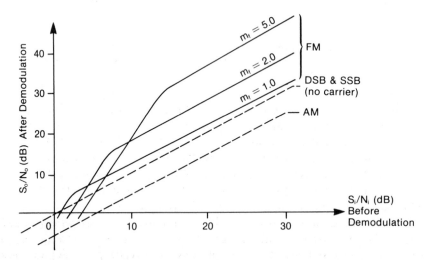

Figure 6.34 Noise performance of various modulation systems. Signal-to-noise ratios before and after demodulation are compared.

The figure shows that:

- Double sideband and single sideband modulation have the same S/N ratio both before and after demodulation.
- Amplitude modulation is worse because of the wasted carrier power.
- FM with modulation indexes less than 1.0 are also worse than DSB and SSB.
- FM with modulation indexes above 1.0 result in significant improvements but only if the input S/N ratio stays high.
- FM (and PM) have a "threshold" problem. When the input S/N ratio falls below the knee on the curve the output signal suddenly becomes much noisier.
- The threshold is unfortunately higher for the wider deviation FM.

The existence of the FM threshold is one of the problems that prevents FM from being a universal modulation system. The other big problem is the large amount of bandwidth required. The threshold can be observed if you are listening to an FM station on your car radio and driving away from the city where the transmitter is located. The music will sound noise-free for a great distance and then, suddenly, in a few short miles (or kilometers) the noise level will build up until the music becomes too hard to listen to.

The threshold is also noticeable when driving around in a city with large buildings. Although the transmitter is close by and the signal should be strong, there will be numerous places where radio waves bounce off buildings causing a weakening and strengthening of the total received signal. A sort of "flutter" may be heard as the signal strength drops below the threshold level. A rapid sequence of flutters that may be noticed as a car moves is called a *picket fence effect*.

FM threshold is the result of random noise voltages disturbing the phase of the FM sine wave. This sine wave can be represented as a vector that rotates at different rates. The different rates are the result of the frequency shifts that represent the information being carried as modulation. The initial vector will have a constant length since the amplitude of the FM signal is always a constant. Noise, on the other hand, has a random phase and a random vector length. The "average" length of the vector is roughly the equivalent of its rms voltage. Figure 6.35 shows what happens when the FM vector and the noise vector combine.

In the first case, the noise amplitude is small so this would represent a situation where the S/N ratio is relatively high. The noise voltage adds to the FM vector at some random angle and produces a new instantaneous amplitude and phase that is likely different from that of the original FM signal. The amplitude change doesn't matter since any demodulator used should be insensitive

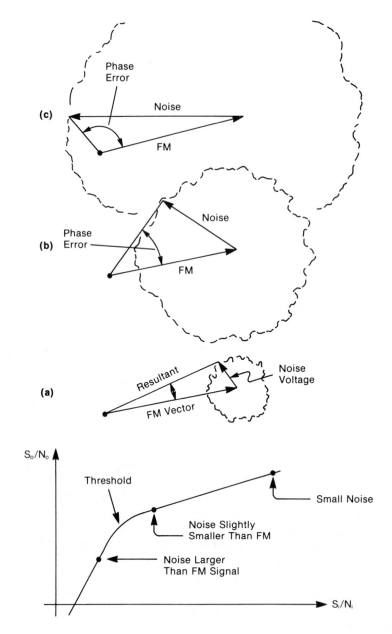

Figure 6.35 FM threshold occurs when the noise voltage has a good probability of being larger than the desired FM signal voltage. Very large phase errors can then occur.

to amplitude changes and, if it isn't, a limiter should be used. The phase change is the serious part since, as we have seen, phase changes and frequency are related.

With the small amount of noise, the diagram shows that the phase change will be quite small and a little math would show that the angle change is proportional to the noise amplitude. This represents the slow decrease in output S/N ratio as the input signal gets noisier.

When the noise amplitude becomes much larger and is about equal to the FM signal's amplitude, much more drastic phase changes occur. Small increases in the already large noise will cause very large increases in the phase angle. When the noise becomes only slightly larger than the carrier, complete phase reversals are possible which could lead to full 360° phase jumps of the rotating vector. The noise is now causing a severe distortion of the FM signal and this is represented by the rapid drop in S/N ratio below the knee or threshold of the curve.

6.13 REVIEW QUESTIONS

1. What is one benefit of FM over AM?

2. What is one disadvantage of FM over AM?

3. What is an advantage of PM over FM?

4. What is the modulation index when a 600 kHz carrier is deviated between peaks of 550 and 650 kHz by an 8 kHz sine wave?

5. Using the Bessel coefficients in Figure 6.5, determine the frequency and amplitude of each sideband component that results when a 100 kHz 5 volt carrier is deviated to peaks of $\pm$ 16 kHz by a 4 kHz sine wave.

6. What approximate bandwidth is needed for the FM signal of question 5?

7. What capacitor value must be used in the de-emphasis network of an FM broadcast receiver if the associated resistor is 10 k ohms?

8. What happens if the 90° phase shift network used in an Armstrong phase modulator shifts the carrier by 85° instead of 90°?

9. What is the total bandwidth needed by the FM signal at the output of a frequency tripler if the input signal is deviated to peaks of $\pm$ 8 kHz by a 5 kHz sine wave?

10. Compared to an AM signal, how much quieter (in dB) is an above-threshold FM signal after demodulation if it consists of a 3.5 kHz sine wave modulating a carrier to a peak deviation of $\pm$ 7.0 kHz? (See Figure 6.34.)

7. TRANSMITTERS

Transmitters are at the "sending end" of a communications system and receivers are at the other end. In between is the noisy, lossy media through which the transmitted signal must travel—suffering untold misfortunes along the way.

The transmitter's characteristics must be selected with this view in mind. It must send out enough power to provide an adequate signal-to-noise ratio at the receiver. It must use the correct modulation to protect the information being carried from excessive distortions. Its frequency must be chosen according to the available channels and the desired area of coverage.

7.1 LEGAL REQUIREMENTS

Some important considerations revolving around the use of transmitters are legal aspects. If not designed and adjusted properly, a transmitter could seriously interfere with other, sometimes vital, communication services. For this reason:

- Channel users are licensed by a Federal controlling agency within each country and this agency, in turn, must work with the International Telecommunication Union (ITU) to ensure that adjacent countries use compatible frequencies.

- Transmitter power output levels will be limited to control their area of coverage and reduce the chance of causing interference.

- Bandwidths used by each transmitter are carefully limited to reduce interference to users of adjacent channels.

- Operators and maintenance personnel must be tested and licensed to ensure that they are aware of these requirements and that they have the ability to keep the equipment operating within the required limits.

7.2 BASIC TRANSMITTER

Most radio frequency (RF) transmitters, regardless of the type of modulation used, would have a block diagram that is roughly the equivalent of that shown in Figure 7.1. This will serve as an introduction to the basic transmitter and some of the terminology used. More specific details of an AM transmitter will be discussed later.

The starting point is the oscillator circuit that ultimately determines the final operating frequency for the transmitted signal. Normally, this would be based on the resonant frequency of a quartz crystal because it is here, in the oscillator, that the overall frequency stability of the transmitter is determined. A typical legal requirement might be that the final radiated signal must remain within .005% of its assigned operating frequency. A crystal is the only reference that is stable enough to meet this requirement at a realistic cost. Some navigation transmitters require much greater frequency accuracy and stability and so use very expensive atomic standards.

Typical power levels at the output of the oscillator circuit will be limited to a few milliwatts. As mentioned in the chapter on oscillators, this causes fewer heating problems and so promotes greater stability. It does, however, mean that considerable amounts of power gain will be needed to raise this to the final level needed in most transmitters.

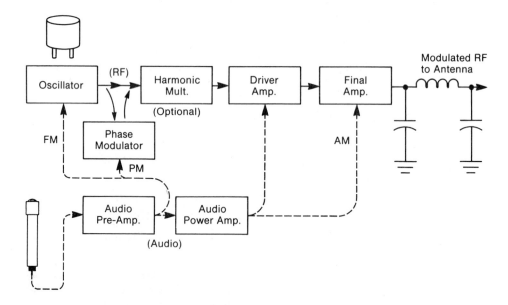

Figure 7.1 General block diagram of a transmitter. The point at which the modulation is added depends on the transmitter type.

For an FM or PM transmitter, the modulating signal would go either directly to the oscillator circuit or to a phase shift network immediately after it. In either case, the audio power level would not need to be very high since the RF power level at the point of modulation would be low.

For transmitters working below 20 MHz, the crystal could be selected to work at the final output frequency. However, for higher frequencies, this would change. Crystals are most stable if kept below this 20 MHz limit. To reach higher frequencies, harmonic multipliers can be used. These are simply amplifiers that are intentionally nonlinear so that their output signals will contain harmonics of the input signal. A tuned circuit can then be placed at the output to select the desired harmonic—usually two, three, four, or five times the input frequency. The multiplier output frequency will have the same percentage stability as the input did. Consider the following example.

Example 7.1

The input to a frequency tripler is supposed to be 1.000 MHz. However, the input wanders around by ±0.015%. What happens to the output?

Solution:

At the input, this percentage drift would result in an error of:

$$\frac{.015}{100} \times 1.000 \text{ MHz} = 150 \text{ Hz}$$

The output frequency will be exactly three times that of the input. For a nominal input of 1.000 MHz, the output will be:

$$F_{out} = 3 \times F_{in}$$

$$= 3 \times 1.000 \text{ MHz}$$

$$= 3.000 \text{ MHz}$$

This output will wander around by the same 0.015% that the input did but now the frequency error will be:

$$\frac{.015}{100} \times 3.000 \text{ MHz} = 450 \text{ Hz}$$

The result is a greater absolute error. At higher operating frequencies it is, therefore, harder to maintain absolute accuracy.

The next block in the diagram is the *driver*. This circuit simply increases the power level of the RF signal without changing its frequency. Several driver stages may be needed, depending on the final power level that must be reached and the frequency that they operate at. One problem is that, as the operating frequency and the power level increase, it becomes harder to obtain high gain with a transistor (or vacuum tube). Larger, higher-frequency transmitters may have many driver stages with the design of each stage changing as higher and higher powers are being handled.

Drivers are usually placed after any frequency multiplier stages because a driver is normally more efficient than a multiplier in terms of RF power out for DC power coming in. If, for example, a multiplier were only 30% efficient, it is better to waste the remaining 70% when the power level is down at 100 milliwatts than up at 10 watts.

For an AM transmitter, the driver stages near the final output point may be amplitude-modulated although not normally to a full 100%. This helps to provide more drive power to the final stage when it is being modulated to higher power peaks and reduces the drive when the final is reducing its power.

The final stage isn't really any different in principle from any of the previous driver stages. The main feature is its higher power level and the fact that it will bear the consequence of any problems with transmission lines and antennas connected to its output. For this reason, there will be several extra adjustment and measurement points associated with the final stage. There may also be some automated fault detection included to prevent the total destruction of a set of expensive transistors should any problems arise.

For an AM transmitter, the power supply to the final will likely be modulated so that the power supply voltage will swing up and down in step with the modulating signal. This will require an added audio power amplifier with a power output at least equal to one-half the DC power input to the final stage. Because of this added audio voltage, the final RF stage must handle voltages considerably higher than the rest of the transmitter. This could be harder on the transistors used and it could be harder on the unsuspecting technician who accidentally pokes a finger into these circuits. With vacuum tube circuits operating at high powers, these voltages are LETHAL even when a finger is held several centimeters away. (An acquaintence of the author was killed doing exactly this—don't let it happen to you!)

Final amplifiers for FM and PM transmitters operate with a fixed-supply voltage, making the design and adjustment of these circuits easier.

The last component in the block diagram is a network that acts both as a low-pass filter and also an impedance-matching network. The filtering is necessary to remove unwanted harmonics of the operating frequency that would otherwise interfere with users of other channels. The impedance matching transforms or changes the impedance of the load (antenna and transmission line) up or down to the optimum value needed to run the final stage at the desired power level and with the greatest efficiency possible.

7.3 AM TRANSMITTER EXAMPLE

The material on transmitters so far has been very general. Now we examine an actual circuit in detail. Figure 7.2 represents a low-power AM transmitter such as would be used for Citizen's Band operation.

7.3.1 Oscillator

The oscillator transistor is biased at about 5 mA collector current and acts as a linear amplifier with positive feedback. Its configuration is similar to the Clapp oscillator circuit shown in Figure 4.12. For Citizen's Band operation (27 MHz), the crystal would likely be a third overtone type. This means that the crystal will be physically vibrating at three times its fundamental frequency. (The thickness of a crystal determines its fundamental frequency.) The result is a quartz wafer that isn't as thin and delicate as it would have been if a 27 Mhz fundamental had been used. The collector-tuned circuit becomes necessary to ensure that the circuit does operate at the third harmonic instead of some other harmonic.

The positive feedback path from emitter to base consists of capacitors C_2 and C_3 and the crystal itself—operating between the series and the parallel resonance points where it has an inductive impedance. The variable capacitor in series with the crystal plays two roles. First, it keeps DC voltages away from the quartz wafer. Although not absolutely necessary, remember that the wafer is stressed and bends proportional to any applied voltage. The removal of any DC component keeps the crystal operating within its linear limits. The second purpose of the variable is to allow some small adjustment in the operating frequency of the oscillator.

The power supply to the circuit is provided with some extra filtering in the positive power line to keep the voltage as smooth as possible. Excessive supply ripple could cause operating frequency shifts and so result in unwanted FM modulation.

7.3.2 Driver

The second stage of the transmitter is a power amplifier. Its sole purpose is to provide enough gain to raise the 5 milliwatt output from the oscillator up to about 150 milliwatts, which would be needed to drive the final transistor. The required gain of 14.8 dB is easily obtained at this frequency and power level with one transistor. (For higher frequencies and powers it becomes increasingly more difficult.)

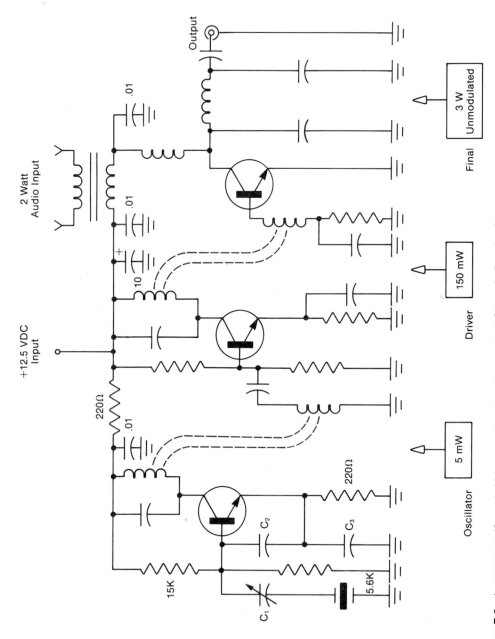

Figure 7.2 Low-power (3 watt) AM transmitter circuit typical of Citizen's Band units. Modulation is supplied by an external 2 watt audio amplifier.

The driver transistor is initially biased at several milliamps of collector current. This is the bias point provided by the voltage divider at the base of the transistor without any signal coming in from the oscillator. When the oscillator is running, the average current through the driver stage will climb to about 20 mA. This is because the base signal is relatively large compared to the bias point. The positive peaks of the sine wave will cause an increase in current but the negative peaks won't cause a corresponding decrease since the transistor will cut off (Figure 7.3) The amplifier, as a result, isn't a linear amplifier because the current flowing through the transistor won't be a sine wave. Harmonics will be generated. However, these harmonics will be at multiples of the 27 MHz operating frequency and so tuned-circuits after the driver stage can easily re-move the unwanted signals, leaving only the fundamental sine wave.

An important point to be remembered here is that, for a nonlinear ampli-fier such as this, the average current depends on the strength of the input sig-nal. This is handy to use when troubleshooting a defective transmitter or adjusting a good one. For example, if the oscillator in this circuit wasn't work-ing, the bias current in the driver stage would drop back to 2 or 3 mA instead of the more normal 20 mA.

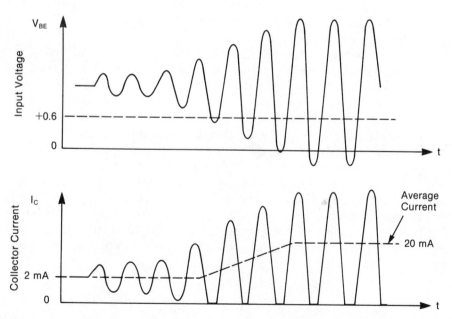

Figure 7.3 When the input signal to the base of a transistor becomes large enough to cause cutoff on the lower peaks, the average bias current will increase.

7.3.3 Final Stage

(You may want to refer back to Chapter 5 on Amplitude Modulation and to Figures 5.12 and 5.13 before you finish this section.)

The final stage is another power amplifier but this one has a variable gain. A 2 watt audio amplifier and audio transformer supply enough extra energy to cause the power supply voltage to swing between 0 volts and twice the normal 12.5 volt DC supply. The final transistor must keep operating between cutoff and saturation at the RF rate for this full range of varying supply voltages.

When the effective supply voltage is 12.5 volts, the RF output will be about 3.0 watts. For 150 mW input from the driver, this would represent a gain of 13 dB.

For the short time that the total supply voltage is at its 25 volt maximum, the output power will be 12 watts. (Doubling the voltage produces four times the power.) The final transistor's gain under this condition will be 19 dB.

At the other extreme, when the power supply swings down to 0 volts, the RF output power will also be 0 and the transistor's gain will be an infinite loss—at least in theory. (With any practical transistor, some of the input signal sneaks through internal capacitance to provide a small output.)

Averaged over a full cycle of the audio waveform, the average RF power will be 4.5 watts. Note that this is not simply the average of the highest power (12 watts) and the lowest power (0 watts). The average output power is 1.5 watts higher with full modulation than it was without modulation. The extra energy was supplied by the audio amplifier that had to be rated at about 2 watts to make up for losses in the final transistor.

The collector breakdown voltage of this transistor must be 50 volts. This will accommodate the peak 25 volts on the modulated supply line plus the fact that the RF signal at the collector with a tuned circuit load can swing up to twice the instantaneous supply voltage. On a transistor data sheet, this breakdown voltage will likely be listed as V_{cer} (maximum collector voltage from collector to emitter with the base voltage reversed—transistor cutoff.)

The circuit shows the final transistor operating without any forward bias on its base. Collector current can only flow when the RF signal at the base rises above + 0.6 volts and, since this will only happen for less than half of each cycle, the biasing is called class-C. The resistor and capacitor, in series with the base lead, maintain an average negative voltage on the base with a value dependent on the strength of the incoming signal from the driver stage. This keeps the positive peaks of the base voltage always clamped at about +0.7 volts.

7.3.4 Output Network

The last portion of our simple AM transmitter is a combination filter and impedance-matching network. A three-element, low-pass filter is provided to reduce

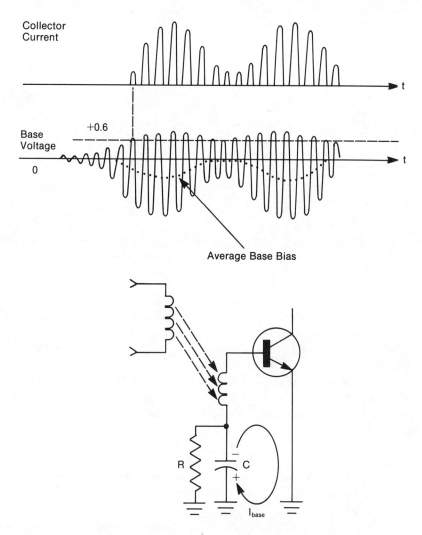

Figure 7.4 Class-C base bias. Collector current flows only when the RF signal drives the base above 0.6 volts. The base current that also flows will build up a negative bias on the capacitor.

the amplitude of harmonics that are the result of the nonlinear operation (cut-off and saturation) in the class-C final transistor. The three reactive components are capable of an attenuation slope of at least 18 dB per octave in the initial part of the stopband. This will reduce any second harmonic by at least 18 dB and any third harmonic by at least 28 dB. Whether this is adequate would depend on the original harmonic amplitudes before filtering and the legal requirements that govern the frequencies and power levels used.

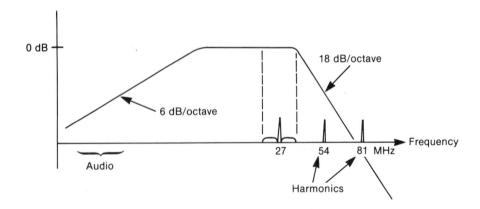

Figure 7.5 The final network provides a filtering function to reduce the amplitude of unwanted harmonics and an audio component. Operation at 27 MHz is assumed here.

The network also provides a single series capacitor for high-pass filtering. This removes the DC and audio components that still exist after the amplitude modulation is created. Figure 7.5 shows the relative placement of the desired and undesired signals and the filter's response.

The third function of the network is impedance matching. It must transform the impedance of the load (usually an antenna) into the impedance needed for the output transistor. These impedance levels will be described in the next section.

7.4 IMPEDANCE LEVELS AND MATCHING

Between the stages and at the output of our sample transmitter, several forms of coupling networks were included. Their function is to transfer as much RF power as possible from one stage to the next and to accommodate changing impedance levels between the output of one stage and the input of the next. What determines these impedances?

One of these is the input impedance of the transistor connected to the output of the coupling network. This transistor will generally have a low input resistance and it will also have a significant reactive component—usually capacitive. The impedance will vary depending on the exact transistor used, the operating frequency, and the power level. Normally, the impedance value would have to be read off a graph in a data book.

Figure 7.6 shows how the two parallel input components would appear for a typical medium-power RF transistor. Notice that the input resistive compo-

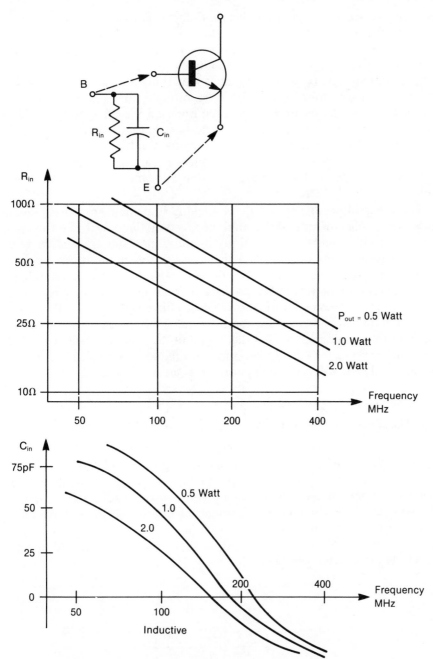

Figure 7.6 Parallel input resistance and reactance of an RF power transistor. The components depend on operating frequency and power level. At lower frequencies, the reactive component is capacitive; at higher frequencies, it may become inductive.

nent decreases with frequency and also with increasing output power level. The parallel reactive portion is capacitive at the lower frequencies and becomes less so as frequency increases and the transistor loses gain. At very high frequencies, the reactive component becomes inductive.

The student may be wondering how something as nonlinear as the base-emitter junction of a transistor, which is operating a small part of the time forward biased and a larger part of the time reversed biased, could be assigned equivalent linear component values. The answer is that harmonics are generated, of course, by the junction diode but only the fundamental sine wave components are considered when the impedance values are calculated. This simulates the single operating frequency that appears in the final transmitter circuit.

The other impedance needed for making impedance-matching calculations is at the collector side of the transistor feeding the power into the matching network. It would be natural to assume that the transistors output impedance would be read off a data sheet and this value used for the design. However, this isn't done. A transmitter design isn't a design for maximum gain; it requires certain amounts of RF power at certain points in the circuit and it wants this power obtained with the greatest efficiency possible. If transistor input and output impedances are carefully matched, the gain will be high but also 50% of the power generated at the output of the transistor will be wasted in the internal resistance. (This may sound contradictory but isn't.) For transmitter design, only the following two parameters are considered:

- The power supply voltage.
- The desired RF power level.

When the DC collector current is fed through the inductor of a collector-tuned circuit, the maximum peak-to-peak collector swing can be almost twice the power supply voltage. A nominal 1.0 volt is usually subtracted from each peak since, at higher frequencies, a transistor doesn't normally have enough time to move deeply into saturation.

Using this maximum value, the power that would be developed in a given value of load resistance would be:

$$P_{load} = \frac{(V_{rms})^2}{R_{load}}$$

$$= \frac{[0.707(V_{cc} - 1.0)]^2}{R_{load}}$$

$$= \frac{(V_{cc} - 1.0)^2}{2\,R_{load}}$$

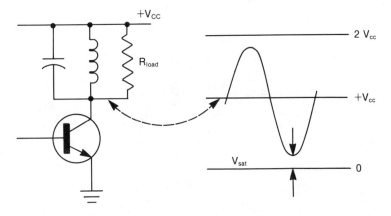

Figure 7.7 The maximum output power from a transistor depends on the power supply voltage and the load resistance. The peak-to-peak collector voltage swing can be almost twice the power supply voltage when a tuned circuit is used.

To find the value of load resistance needed to obtain a certain required RF power, simply reverse the formula:

$$R_{\text{load}} = \frac{(V_{cc} - 1.0)^2}{2 \, P_{\text{load}}}$$

Just placing this value of resistance at the collector isn't a guarantee that the desired power will magically pop out of the transistor. First, the transistor must be capable of handling the desired power without destroying itself. Second, it must have sufficient gain to amplify the small input signal at the base up to the desired power level. The transistor data sheet will indicate whether a device is adequate for the job intended. In fact, most manufacturers will recommend specific transistors for operating frequency, supply voltage, and power levels.

Example 7.2

Find the turns ratio (n_p/n_s) necessary to develop 1.5 watts at the output of the first transistor and pass it on to the input of the second. Assume R_{in} of the second transistor is 7.5 ohms.

Solution:

With a 10 volt power supply and a required output of 1.5 watts, the necessary impedance for the collector of the first stage will be:

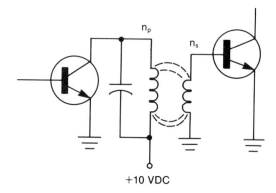

+10 VDC

Figure 7.8

$$R_{\text{load}} = \frac{(V_{cc} - 1.0)^2}{2\,P} = \frac{(10 - 1)^2}{2(1.5)}$$

$$= \frac{9^2}{3} = 27\Omega$$

The impedance ratio between the primary and the secondary is 27Ω to 7.5Ω.

The turns ratio is the square root of this; therefore:

$$\frac{n_p}{n_s} = \sqrt{\frac{27}{7.5}} = 1.90$$

(This assumes that complete coupling exists between primary and secondary.)

7.5 RF POWER TRANSISTORS

One of the more interesting features of a solid-state transmitter is the construction of the high-powered transistors used as driver and final amplifiers. This will be discussed first, since many of the resulting characteristics control the design procedure to be used. Power amplifiers must handle large currents, and since current density in a transistor must be kept at reasonable levels, a large chip area inside the transistor is called for. Unfortunately, this area could result in very large junction capacitances that would very seriously limit the useful gain and bandwidth of the device. But all this capacitance is not necessary. Power transistors, particularly at higher frequencies, where gain is low, will have fairly large base currents. As shown in Figure 7.9, the base current flowing through the resistance of the base region will produce a lateral voltage drop. The highest base-to-emitter bias will be developed at the edge of the emitter region, and so the emitter current will be strongest at that point. The area directly under the emitter region will have a lower bias, so will contribute very little to the transistor action, only unwanted capacitance.

The practical solution is to make the horizontal dimension as small as possible; in fact, one figure of merit used in the design of power transistors is the periphery-to-area ratio, which should be kept high (8:1). As both operating frequency and power levels increase, the *P/A* ratio becomes increasingly critical, and the top view of the transistor shows a series of long narrow fingers for both the base and emitter metal contacts. Each emitter finger has two edges that concentrate the emitter current in the space down between the adjacent base fingers. The limit on the narrowness of these fingers is set by processing problems and by the minimum amount of metal required to handle the current levels without squirming or buckling. A view of a typical power transistor chip is shown in Figure 7.10.

With all the long fingers, it becomes very important to ensure that each portion of the transistor shares the current equally. Any tendency for current to concentrate in one area will increase the temperature at that point, resulting in lower resistance and the possibility of thermal runaway. The resulting "hot spot" could damage the transistor if the temperature climbed high enough. The emitter current can be shared more equally and thermal runaway prevented if small resistors are included at each emitter. A schematic representation of this and a diagram are shown in Figure 7.11.

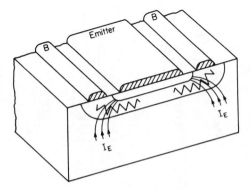

Figure 7.9 Cross section of an RF power transistor showing the crowding of emitter current near the emitter edges.

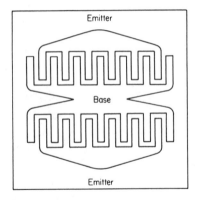

Figure 7.10 Top view of an RF power transistor with narrow emitter and base fingers to maximize the edges and minimize emitter area.

7.5.1 Packaging

When a good power transistor chip is produced, the job is only partially finished. The chip must then be packaged for final use. The two main considerations are the removal of heat from the chip and the electrical connections to the emitter, base, and collector.

For best heat removal, the collector can be connected electrically to the case, or if this is undesirable, the collector can be brought out as a separate

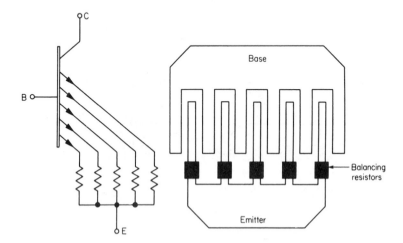

Figure 7.11 Small resistors can be added to the multiple emitters to encourage more-equal distribution of current and more reliable operation.

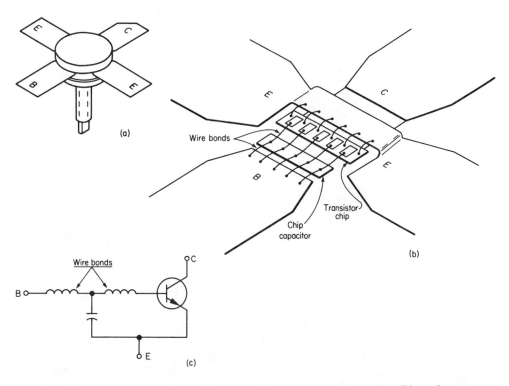

Figure 7.12 Exterior view of an RF power transistor (a), its internal view (b), and equivalent circuit (c).

lead. For minimum inductance and ease of connecting to printed circuit boards, the leads are usually wide metal tabs and the emitter is often a double tab.

The exterior and interior view of a popular package are shown in Figure 7.12. The interior view shows a matching network, consisting of a ceramic chip capacitor and controlled wire bonds, built between the internal base and the external base lead. This is done in some transistors for the designer's convenience, to raise the very low input resistance of the transistor (about 1 Ω or less) up to something a bit more manageable (several ohms) over a limited range of frequencies (less than one octave).

7.6 TRANSMITTER MEASUREMENTS

During the design phase of a new transmitter or when troubleshooting and maintaining an older one, measurements have to be made. These are compared to manufacturers' specifications and Federal requirements to determine if the circuitry is operating legally and efficiently. The more important measurements consist of the following:

- Operating frequency
- Stray signal check
- Modulation
- DC power input
- RF power output

During any tests and adjustments, the transmitter should not be operated "on the air"—in other words, it should not be connected to an antenna. Otherwise, there would be a danger of causing interference to other radio users if the wrong adjustments were made. Instead *dummy loads* are used. These consist simply of resistors that have the same value as the antenna impedance that the transmitter was designed to work with. The full RF power from the output will now be heating these resistors so they must be selected to safely dissipate the power without overheating. For a 5 watt transmitter, this is a simple requirement; but, for a 10 kW transmitter, an elaborate water-cooled unit would be needed.

Never try to operate a transmitter without either a dummy load or an antenna. The chance of destroying one or more of the high power transistors within a few milliseconds is very good. The unfortunate part is that with lower-power circuitry you don't even hear a little "poof" as the transistor melts. Higher-power circuitry is much more obliging.

7.6.1 Frequency Check

The operating frequency of a transmitter is best checked using a good frequency counter connected after the dummy load. Any modulation should be turned off so that only a single frequency (for AM and FM transmitters) is produced—the carrier. SSB transmitters pose a special problem for, without modulation, there would be no output at all. In this case, an audio tone (possibly 1000 Hz) could be used for modulation and one sideband would appear. This would be 1000 Hz higher or lower than the suppressed carrier depending on which sideband was selected.

The frequency counter must be a good quality instrument that is regularly checked against an accurate frequency standard. Cheap or poorly maintained counters can easily display an impressive number of digits without being accurate. Keep in mind that this measurement may have to say whether the transmitter is operating legally within its assigned band or illegally outside it.

In some cases, the counter should not be connected directly across the dummy load; the RF voltage could be too high for the counter input stages if the transmitter power level is large. For example, if the carrier output power is 100 watts, the RF voltage across 50 ohms would be 70.7 volts rms. This could be too high for some counters and alright for others—check the counter's operating manual. Figure 7.13 shows a 20 dB attenuator "pad" connected between the dummy load and the counter input. (The 20 dB value printed on the side of the attenuator used will only be correct if the counter input resistance is also 50 ohms; although, for a frequency measurement that isn't too serious.)

If the frequency measurement shows that the transmitter is off frequency by a small amount (less than 1 or 2%), the problem can be corrected by adjusting tuning elements within the oscillator circuit. If the error is larger than this, a serious problem could exist and a much more detailed look at the equipment will be necessary.

7.6.2 Harmonics

A faulty transmitter could cause unwanted signals to appear on frequencies other than the intended one. These could be one or more harmonics, sub-harmonics (if frequency multipliers are used), or other nonrelated frequencies (spurious signals). The modulation should be turned on for these checks just in case it plays some part in the generation of these unwanted signals.

A wide-range receiver with a calibrated signal strength meter could be used to tune across the necessary frequency range but this could be quite tedious, especially if the procedure needs to be repeated frequently while adjustments are being made.

A better and easier check can be made using a spectrum analyzer connected across the dummy load (with extra attenuators if necessary). The ana-

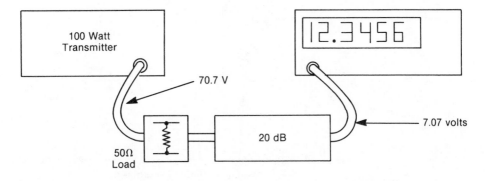

Figure 7.13 A 20 dB attenuation pad is used to reduce the high RF voltage from the transmitter down to a safer level for the frequency counter.

lyzer would be set for a wide frequency display and the carrier, sidebands, and any unwanted extra signals would appear as vertical lines on the display. With all signals visible at the same time, it is easier to make adjustments and harder to miss a weaker but still significant signal.

Using either the receiver or the spectrum analyzer, frequencies from the lowest used in the internal circuitry of the transmitter up to at least five times the normal output should be checked. The results should then be plotted on a graph of signal level versus frequency. The signal level is usually plotted in decibels but this could be either in absolute power level (dBm) or in dB relative to the main carrier power. These final measurements must include any attenuation placed between the dummy load and the measuring device. Figure 7.14 shows a typical result along with a fictitious legal limit which indicates that the second harmonic is too strong.

If the harmonics are too strong, what do you do? In some cases, resonant circuits called *traps* are included in the transmitter design to reduce the strength of specific frequencies. These can usually be adjusted to obtain the desired results without seriously affecting the overall operation of the transmitter.

The main task of harmonic reduction belongs to the low-pass filter section after the final transistor(s). But, this isn't the only function performed by this circuitry; it must also provide the correct impedance matching between the antenna and its transmission line and the collector side of the final transistor(s). It may be possible to make small improvements in harmonic reduction by re-tuning this section but it must also be kept in mind that power level and overall efficiency could suffer at the same time.

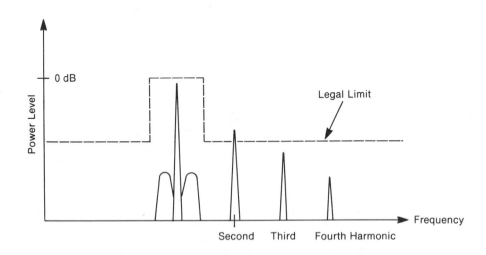

FIGURE 7.14 Total signal spectrum at the output of a transmitter.

If everything else appears to be OK, an "on-the-air" check could be made to look for harmonics. Again, a calibrated receiver would be needed and in this case it would be connected to a separate antenna located some distance away from the transmitter's antenna. The results of this test could be slightly different from those obtained with the dummy load. The reason is that the impedance of an antenna won't be the same at the harmonic frequencies as it is at its normal operating frequency. This will upset the attenuation characteristics of the low-pass filtering.

If all measurements are completed and the harmonics are still too strong, extra filtering will be needed. A separate low-pass filter can be added in the transmission line between the transmitter and the antenna. This will help with the harmonics but do nothing for subharmonics or any spurious signals immediately above the main operating frequency.

7.6.3 Modulation

Some type of check should be made to determine how well the modulation circuitry of a transmitter is working. The measurements would have to indicate how strong the modulation is (i.e., percent modulation) and how badly distorted the results are when compared to the original modulating signal. The equipment needed will depend on the type of modulation being checked.

For AM, a good high-frequency oscilloscope may be sufficient. If its frequency response is adequate, the RF waveform at the dummy load can be observed while single audio frequencies are used to modulate the transmitter. From the pattern, the percentage modulation can be determined as shown in Figure 7.15. A visual check for distortion can be made at the same time by comparing the original modulating signal with the RF envelope. Overmodulation that can cause excessive sidebands can also be detected this way.

Accurate percentage distortion measurements require the RF waveform to be carefully demodulated first. Then the single original modulating frequency is filtered out with some type of narrow-band "notch" filter. Any remaining signals will then represent unwanted distortion that has occurred during modulation and these can be measured and expressed as a percentage of the original signal.

Distortion can also be measured using a spectrum analyzer. Anything beyond the first pair of sidebands (assuming only a single modulating frequency was used) will represent distortion products and their amplitudes can be measured.

SSB transmitters can be checked in much the same way. The main difference is that two audio tones will be needed at the same time and the resulting envelope will represent the difference frequency.

For FM and PM transmitters, no modulation envelope will be seen on an oscilloscope or at least there shouldn't be. If there is, it means that some form

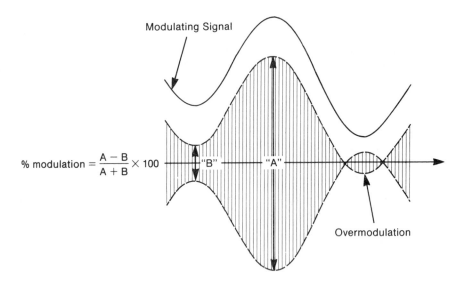

Figure 7.15 AM envelope can show percentage modulation, distortion, and overmodulation.

of AM is accidentally occurring and must be corrected. The spectrum analyzer is the more valuable tool for general testing. With it, the upper and lower sidebands can be checked to see if they are symmetrical. If they aren't, FM and PM could be occurring simultaneously and this should be corrected. Frequency deviation can also be measured from the displayed spectrum. Distortion measurements are too difficult to obtain from the spectrum since, even for a single modulating frequency, too many sidebands are generated. Any unwanted ones would be difficult to see. Demodulation is the best answer and again it must be done carefully so that extra distortion isn't added in the process.

7.6.4 DC Power Input

In some cases, the legal maximum "power" of a transmitter is defined in terms of the DC power fed to the final RF amplifier(s) from the power supply. This won't be the same as the RF output power (the RF power will generally be about 40% lower) but it is easy to measure, requiring only a DC voltmeter and ammeter. Apart from the legal requirements, it is often worth the effort to measure this power and use it to make an efficiency calculation of the final stage.

$$\text{DC input power} = V_c \text{ (average)} \times I_c \text{ (average)}$$

DC input power is calculated by multiplying the average power supply voltage times the average collector current drawn by the final stage. The point at which the current is measured must be selected carefully so that no driver-stage current is included. If current to multiple transistors is being measured, these should only be transistors that directly contribute to the final output power—for example, parallel output transistors. The measurement for the final of a small AM transmitter is shown in Figure 7.16.

By using the average collector voltage, any variation in supply voltage due to amplitude modulation of the supply is eliminated. The resulting power calculation represents only the power supply energy consumed by the carrier. During modulation, the total energy input is higher due to the added audio power.

For efficiency calculations, the RF output power is measured (with no modulation) and then compared to the DC input power.

$$\text{Final stage efficiency} = \frac{\text{RF output power}}{\text{DC input power}} \times 100\%$$

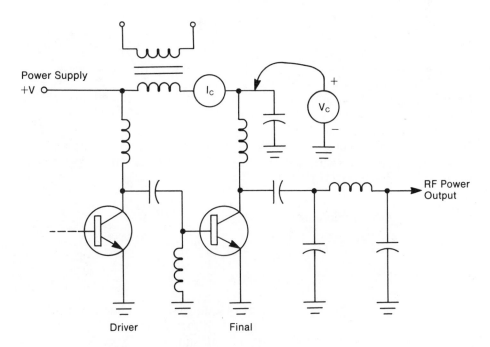

Figure 7.16 DC power input measurement using the average collector voltage and the average collector current.

Example 7.3 ━━━━━━━━━━━━━━━━━━━━━━━━━━━━━━━━━━━━━━

An AM transmitter operates from a 24 volt DC power supply. The modulating signal makes this swing up and down by 20 volts peak. The average current to the final is 150 mA. The RF output power measured without modulation is 2.0 Watts.

What is:

(a) The DC input power.

(b) Stage efficiency.

(c) Total RF output power with modulation.

(d) Transistor heating power (with mod.)?

Solution:

The DC power input to the final stage is:

$$\text{DC } P_{in} = V_c \times I_c$$
$$= 24 \text{ volts} \times 0.15 \text{ amps}$$
$$= 3.6 \text{ watts}$$

$$\text{Final stage efficiency} = \frac{2.0}{3.6} \times 100\%$$

$$= 55.5\%$$

To find the total RF output power with the modulation included, first we have to find how much audio power is coming from the modulator.

We know that the audio voltage swings ± 20 volts peak which is 20/24 ths of the full supply voltage. The audio current will swing up and down by the same ratio. (For a constant load, the current is always directly proportional to the applied voltage—OHM.)

$$\text{Peak audio current} = \text{DC current} \times {}^{20}/_{24}$$

$$= 0.15\text{A} \times {}^{20}/_{24}$$

$$= 0.125 \text{ amps}$$

$$\text{The extra Audio input power} = \text{RMS voltage} \times \text{RMS current}$$
$$= (20\text{v} \times 0.707) \times (0.125\text{A} \times 0.707)$$
$$= 1.25 \text{ watts input}$$

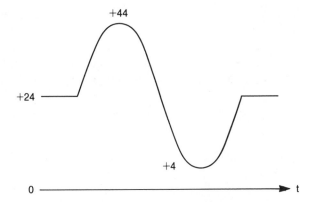

Figure 7.17 Power supply voltage swing due to 24 volt DC supply plus ±20 volts peak audio.

Extra sideband power output
$$= 1.25 \text{ W} \times 55.5\%$$
$$= 0.694 \text{ watts}$$

Total RF output power with modulation
$$= 0.694 \text{ W} + 2.0 \text{ W}$$
$$= 2.694 \text{ watts}$$

Transistor Heating Power:
This is the difference between the total power input to the stage (DC plus Audio) and the RF output. This is what makes the transistor feel hot.

$$\text{Heating Power} = (3.6 + 1.25) - 2.694$$
$$= 2.16 \text{ watts}$$

7.6.5 RF Power Output

Our final check of a transmitter's performance is the measurement of the actual RF power produced at its output terminals. Normally, this would be measured as the power being sent to a dummy load.

The measurement can be made in either of two ways. One possibility is to measure the actual heating power of the RF energy by sensing the change in resistance of a thermistor or the temperature rise of cooling water. The *thermistor* is a very sensitive measurement that can detect power levels well below one milliwatt. It is also commonly used at microwave frequencies.

The second possibility for power measurement is to check the RF voltage across the dummy load and then calculate the power using the formula:

$$P = \frac{V_{rms}^2}{R_{load}}$$

If a voltmeter scale is recalibrated to show the equivalent power, it will have the nonlinear scale shown in Figure 7.18. The common method of measuring RF voltage is to rectify it with a good high-frequency diode and then display the results with a DC meter. The diode may add further nonlinearities to the power scale, depending on the voltage levels being measured. If the RF voltage is large, the 0.6 volt drop across the diode will be insignificant. The speed of the diode can also cause some problems as it means that the calibration may be frequency-sensitive. Calibration should be done at the actual frequencies needed.

With a small capacitor, the meter will respond to the average rectified voltage. As a result, when used with an AM transmitter, any modulation will not change the reading. The meter will only be measuring the carrier power. If the capacitor is made larger, the peak RF voltages will be measured and the meter reading will fluctuate as the modulation changes. (This won't be a true total power because the deflection is based on a peak value instead of an RMS value and also because the scale is nonlinear.)

Power measurements from a SSB transmitter deserve a separate mention. Since there wouldn't be any significant output power when no modulation is present and since the output power level with modulation depends on the audio level, both the measurement and its description will be more complex. SSB

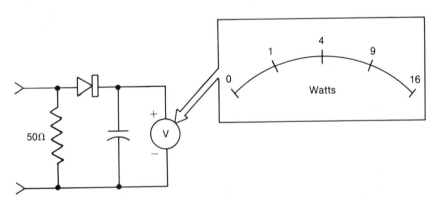

Figure 7.18   An RF watt meter using a rectifier and DC meter. The resulting power scale on the meter will be nonlinear.

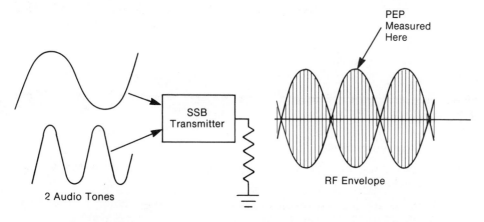

Figure 7.19   Two audio tones are used to modulate an SSB transmitter. The peak
envelope power is the average power during the brief period when the
modulation is at its peak.

transmitters are usually rated in terms of *peak envelope power* capabilities
(PEP). This doesn't imply that peak voltages are used during the power calculations; it means that the power is measured or calculated during the brief time
in the modulation cycle that the true power output is at its highest peak.

It would appear that the simplest way of measuring the PEP would be to
use a strong, single audio tone for modulation and then measure the RF output
power on a normal wattmeter as already described. Since there will be only the
one side frequency produced, there will be no modulation envelope and so the
power will not vary and the wattmeter reading should be the PEP. While this
will work, two practical problems result. First, it will be very difficult to determine how strong the audio should be to prevent "modulation distortion." (With
a single frequency there is no envelope to observe.)

The second problem is that this method will run the transmitter at its full
output power (PEP) continuously—as long as the measurement takes. But few
SSB transmitters are designed to handle this condition. During normal operation, this power level is only reached on modulation peaks and the average
power output would be considerably lower (roughly half).

A safer measurement, therefore, involves the use of two audio tones of
equal strength but different frequencies. Now the RF will show an envelope (at
the difference frequency) and so it can be checked for distortion. The final
stages of the transmitter are also running at their more normal power and so
every one is happier. But, how is PEP measured? Since an oscilloscope will
likely be used to check for distortion, it can also be used to help with the power

calculation. The RF voltage at the highest peak can be determined visually. This can be converted into an rms value and then the power calculated. The other possibility is to read the average output power on a conventional wattmeter and then approximately double it.

7.7 REVIEW QUESTIONS

1. Why must transmitters and their operators be licensed?

2. Why do some situations require high-power transmitters and some low-power?

3. What are the advantages of adding some modulation to the driver stage of an AM transmitter?

4. What are the functions of the output network in a transmitter?

5. Why do high-power RF transistors have an intricate-looking emitter structure with interleaving fingers?

6. Why is a dummy load important for transmitter testing?

7. A class-C RF amplifier operates from a 12 volt power supply and has an RF saturation voltage of 1.5 volts. What is the maximum peak-to-peak collector sine wave?

8. For the same situation as in question 7, an 85 ohm load resistor is connected to the collector through a coupling capacitor. What RF power enters the load? What DC current flows if the transistor is 62% efficient?

9. Using information from Figure 7.6, what is the input impedance (rectangular coordinates) of the transistor when producing an output power of 1.0 watt at 100 MHz?

10. The frequency of a 75 watt transmitter is being checked with a frequency counter and 65 ohm load (Figure 7.13). The maximum input voltage to the counter is labeled as 2.0 volts rms. You have a matching variable attenuator that can be adjusted in 10 dB increments. What should you set the attenuator to?

8. RECEIVERS

A *receiver* is the device that selects a desired narrow range of frequencies that correspond to the output of some particular transmitter. It must then get rid of as much unwanted noise and signals that exist at other frequencies as possible, amplify the weak signal, and then demodulate it to obtain the original intelligence.

8.1 RECEIVER PARAMETERS

There are a number of key parameters that must be kept in mind when receivers are being discussed. These form the basis for a discussion on whether one design is better than another:

1. *Sensitivity.* This is a number that indicates how weak a signal the receiver can pick up and still produce an intelligent signal after demodulation. It is normally measured in microvolts. For example, a shortwave receiver might require 1.5 microvolts at its antenna terminals in order to produce an audio output that is 10 dB stronger than the background noise. Sensitivity and amplifier noise figures go hand in hand.

2. *Selectivity.* This defines the bandwidth that the receiver will pick up and, more importantly, it tells how well it will get rid of unwanted sig-

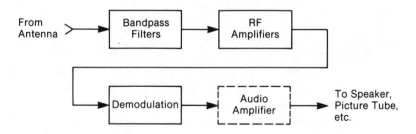

Figure 8.1 Fundamental components of a receiver.

nals that are very close to the desired one. Selectivity is a description of the receiver's filters.

3. *Spurious signals.* Apart from the desired signal and those adjacent in frequency, there may be some other unusual frequencies that unfortunately manage to sneak into the receiver. This specification tells what they are and how serious the problem is. Often this is a problem of either poor filtering or extra mixing in nonlinear amplifiers.

8.2 RECEIVER TYPES

Regardless of the type of demodulation needed, the bulk of a receiver's characteristics are determined by the circuitry ahead of the demodulator section. Several different approaches can be taken to meeting these requirements. All have been tried at one time or another over the years and one has emerged as the dominant design. Keep in mind that, first, we are going to discuss the general topic of receivers without worrying about the type of demodulation that will be used. The demodulation type will require some differences in the main design but these will be examined later.

8.2.1 The Crystal Set

Back in Chapter 1 we saw the most basic of all receivers—the crystal set. It was used by Marconi back in 1901 to detect the weak crackles and hisses from his spark-gap transmitters; and it has been used by many since to listen to near-by AM transmissions. A crystal set, in its simplest form, uses no amplification. Thus, it was the only receiver possible in the early days of radio before vacuum tubes were developed.

The basic circuit, as shown in Figure 8.2, consists of a simple bandpass filter to select the range of frequencies desired and get rid of all the rest. Then a nonlinear device (a diode) is included to do the necessary demodulation. The second capacitor acts as a low-pass filter to remove the remaining high-frequency signals and then the surviving audio can be listened to with the earphones.

This receiver worked fairly well—back in the days when the frequency spectrum wasn't too crowded and listeners didn't mind the high impedance headphones clamped on their ears. Even today, if a listener is close to a strong station without any others near in either frequency or physical distance, the results can be extremely good.

Much of the design of a crystal set must revolve around the characteristics of the diode. If we look closer at the diode's voltage-current curve, we see that there really isn't any abrupt point where the resistance changes from low to

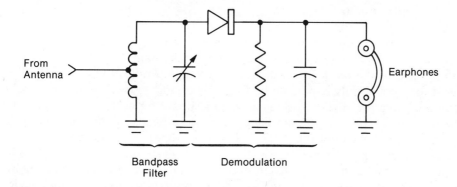

Figure 8.2 Crystal set receiver uses no amplification.

high. (The steeper the slope, the lower the resistance.) Even for very weak signals there will be some change in resistance as the signal strength of an AM signal varies so some demodulation will take place. The problem is that, for weak signals, the diode's changing resistance will all be very high values. For maximum efficiency then, the place where the diode gets its signal (the tuned circuit) and the place where the signal goes (the earphone) must both have impedances slightly higher than that of the diode. The step-up ratio on the tuned circuit and the impedance of the headphones must be chosen with this value and the resulting signal loss in mind.

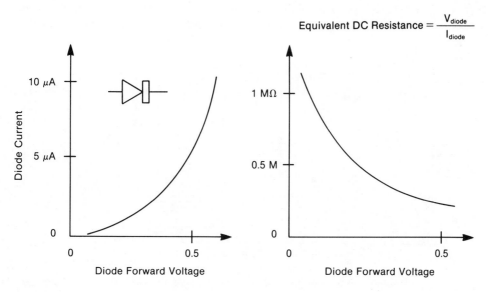

Figure 8.3 The forward resistance of a diode changes smoothly.

8.2.2 Tuned Radio Frequency

The most obvious improvement to the basic crystal set is to add some gain. After all, the received signals can be very weak. Gain can be added either before or after the diode, or both. The first vacuum tubes (and later, the first transistors) could only handle audio frequencies and so the gain could only be achieved after the diode. The result would simply be a "louder crystal set." Then, as better tubes and transistors were developed, radio frequency gain could be added ahead of the diode. What is the difference? One benefit is that the resulting stronger signal makes the diode operate on a steeper part of its slope and so makes it more efficient because its effective resistance will be lowered. A second reason is that any amplifier also acts as an "isolating" device between its input and output circuits. This makes it possible to easily build more elaborate filters without having to worry about tuning interaction. And, it is also possible to separate filter impedances from the diode impedances. A tuned radio frequency (TRF) receiver circuit is shown in Figure 8.4. The result is the tuned radio frequency (TRF) receiver. With more elaborate filtering, it can more easily separate closely spaced radio frequencies. It can pick out the desired station and more effectively reduce the strength of the unwanted ones adjacent to it. This results in better "selectivity" than the crystal set. The added gain ahead of the demodulation circuit will also improve the strength of weak signals and so "sensitivity" should also be improved. However, you might recall from Chapter 1 that each amplifier adds some noise of its own, so sensitivity— the ability to hear weak signals above the noise—will improve only if this added noise is kept to a minimum.

The TRF receiver operates reasonably well. With proper design, it can be quite sensitive. The bulk of all early vacuum tube receiver designs used this

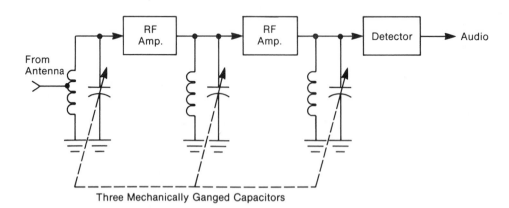

Three Mechanically Ganged Capacitors

Figure 8.4 Tuned radio frequency receiver uses identical tunable circuits separated by radio frequency amplifiers to obtain greater sensitivity and selectivity.

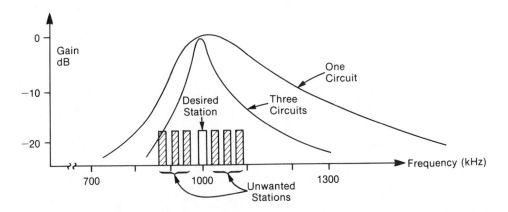

Figure 8.5 Selectivity comparison of the crystal set's single-tuned circuit vs. three identical tuned circuits in the TRF receiver.

design—some with as many as five individual tuning sections. But the TRF design is rarely used today. Why? What is wrong with it?

The main problem is selectivity, or rather, lack of it. The design produces a much sharper bandpass than the single-tuned circuit of the crystal set design but it lacks shape. The peak can actually be too sharp. Since the tuning circuits must all be variable, this makes it nearly impossible to set the bandpass shape and width and keep them constant across the full tuning range.

The shape and bandwidth problems have several causes. Some are readily curable; some are not. A resonant circuit can be tuned by either adjusting its capacitor or its inductor. The capacitor has traditionally been the easier of the two to make variable and, in this case, causes a problem. With a fixed inductor and variable capacitor, the reactance of each element increases as the circuit is tuned to higher frequencies. This results in the loaded Q of the circuit being higher at the low end of the broadcast band (540 kHz) than it is at the other end (1600 kHz).

$$\text{loaded } Q = \frac{R_{\text{parallel}}}{2\pi f L} \tag{8.1}$$

The parallel resistance stays constant as it is mainly due to bias resistors of the adjacent amplifiers. This makes the bandwidth very narrow at the lower frequencies and much broader at the higher frequencies—not the most desirable situation. (Constant would be nice.)

$$\text{BW} = f / Q_{\text{loaded}} \tag{8.2}$$

A better approach is to make the inductors variable and leave the capacitors fixed. Now the reactance of each element decreases as the circuit is tuned to

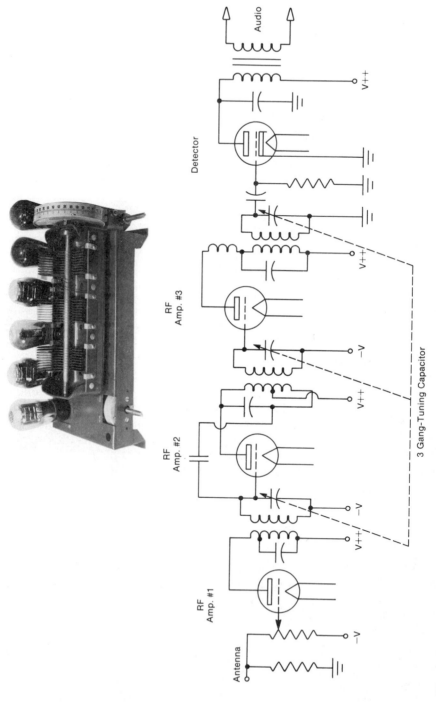

Figure 8.6 Schematic and photo of an old RCA TRF broadcast receiver using vacuum tubes.

higher frequencies. With the fixed parallel resistance, the loaded Q will also increase so the bandwidth will stay reasonably constant. The only problem is that it is harder to make a set of variable inductors that tune together.

$$\text{loaded } Q = R_{\text{parallel}} \times 2\pi f C \qquad (8.3)$$

Now we have a more constant bandpass. What about shape? Multiple-tuned circuits isolated with amplifiers simply produce a sharp peak that is the sum of the individual peaks. No "feedback coupling" takes place.[1] Therefore, it is difficult to create a desirable flat top and steep sides and keep them constant as different frequencies are tuned to.

Example 8.1 ═══

The tuned circuit below must tune from 540 to 1600 kHz. The values given are for resonance at 540 kHz. The inductor itself has no losses and the 15 kΩ parallel resistor is the only loading.

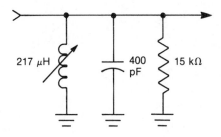

(a) If only the capacitor value were to be decreased so that the circuit would resonate at 1600 kHz, what would the resulting Q and bandwidth be? (The capacitor would change to 45.6pf). The inductor value remains fixed. Therefore:

$$\text{loaded } Q = \frac{R_{\text{parallel}}}{2\pi f L} \quad \ldots\ldots\ldots\ldots\ldots\ldots\ldots \quad (8.1)$$

$$= \frac{15k\Omega}{2\pi 1600 \times 10^3 \times 217 \times 10^{-6}}$$

$$= 6.88$$

[1] If capacitor or inductor coupling is used (see Chapter 2), signals can flow in both directions. The resonant frequencies interact and so flat-topped responses become possible. With amplifier coupling, signals can only flow in the one direction so no interaction occurs.

$$\text{Bandwith} = f/Q_L \quad \ldots \ldots \ldots \ldots \ldots \ldots \quad (8.2)$$

$$= 1600 \times 10^6/6.88$$

$$= 232.7 \text{ kHz}$$

(b) If, instead, only the inductor value were to be decreased so that the circuit would resonate at 1600 kHz, what would the resulting Q and bandwidth be? (The inductor would change to 24.7μH.) The capacitor value remains fixed. Therefore:

$$\text{loaded } Q = R_{\text{parallel}} \times 2\pi fC \quad \ldots \ldots \ldots \ldots \quad (8.3)$$

$$= 15 \times 10^3 \times 2\pi \times 1600 \times 10^3 \times 400 \times 10^{-12}$$

$$= 60.3$$

$$\text{Bandwidth} = f/Q_L \quad \ldots \ldots \ldots \ldots \ldots \quad (8.2)$$

$$= 1600 \times 10/60.3$$

$$= 26.53 \text{ kHz}$$

(c) What is the Q and bandwidth of the original circuit when it is tuned to 540 kHz?

$$\text{loaded } Q = \frac{R_{\text{parallel}}}{2\pi fL}$$

$$= \frac{15\text{k}\Omega}{2\pi 540 \times 10^3 \times 217 \times 10^{-6}}$$

$$= 20.4$$

$$\text{Bandwidth} = f/Q_L$$

$$= \frac{540 \times 10^3}{20.4}$$

$$= 26.53 \text{ kHz}$$

Therefore, variable-inductor tuning keeps the bandwidth constant over the full tuning range.

One final problem (as if we didn't have enough already). The amplifiers in the TRF design must all operate at the frequency being received. Because transistors and vacuum tubes lose gain as their operating frequency increases, this could result in a potential loss of gain and a possible increase in overall noise level.

8.2.3 The Superheterodyne Receiver

If you have been following the discussion so far you will realize that the sensitivity problems have been solved with the TRF design but the selectivity problems haven't been totally eliminated. The superheterodyne receiver corrects this last deficiency. The biggest stumbling block with the TRF design was that all its filters had to be tunable. The superheterodyne (or "superhet" for short) principle puts the bulk of the amplifying and filtering at a lower, fixed frequency called the *intermediate frequency* (IF). Now elaborate filters and high-gain amplifiers can be constructed easily. For AM broadcast band receivers, the standard IF frequency is 455 kHz. For FM broadcast receivers, it is 10.7 MHz.

With the fixed frequency for the amplifiers and filters, the next step is to add a circuit ahead of this to select the proper frequency to be received and move it down to the IF frequency. This step involves a nonlinear mixer plus a variable frequency oscillator. The resulting circuit is shown in block diagram form in Figure 8.7.

Both the mixer and the variable frequency oscillator have been discussed in previous chapters but a little review wouldn't hurt. The mixer circuit should be a nonlinear amplifier with a second-order (square law) curve to its input-output characteristic as explained in Chapters 1 and 5. As such, it will produce, at its output, all the original input frequencies plus their sums and differences taken two at a time.

If a standard IF frequency of 455 kHz is used and the desired frequency to be received is 1.0 MHz, what would the oscillator frequency need to be? There are two possible answers. If the oscillator frequency were 1.455 MHz, the difference between the two inputs would be the correct IF frequency. Another possibility is an oscillator frequency of 545 kHz. Again, the difference will be 455 kHz. Either oscillator frequency could be used but, for AM broadcast work, it is simpler to design tunable oscillators at the higher-frequency side than the lower side of the desired signal.

If we use an oscillator frequency of 1.455 MHz and an intermediate frequency of 455 kHz, are there any other frequencies that could also get through the mixer and be amplified in the IF section? One possibility is a frequency of 455 kHz itself. Another is a frequency 455 kHz above the oscillator frequency—in this case, 1.910 MHz. The latter is known as the *image frequency*.

Now we see why there is a single-tuned circuit between the input and the

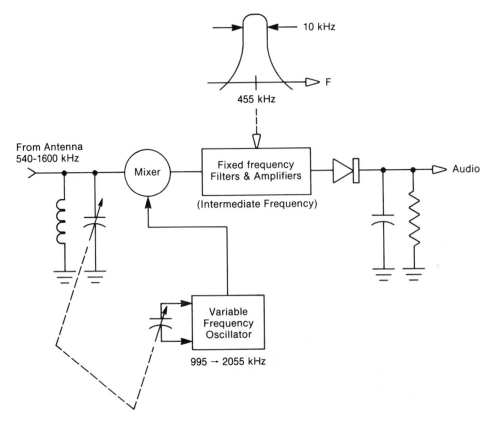

Figure 8.7 Block diagram of a broadcast band AM superheterodyne receiver.

mixer in the block diagram of Figure 8.7. Its job is to eliminate these "other" frequencies that could get into the IF. Notice that this is a different filtering requirement than that performed by the tuning circuits in the crystal set and the TRF designs. In the superhet, this filter does not have to get rid of adjacent channel signals; that is the job of the IF filters. The front-end filter for the superhet simply gets rid of a few troublesome frequencies that are much further away from the desired one and, therefore, are much easier to filter out.

$$F_{image} = F_{desired} + 2\,F_{intermediate\ frequency} \tag{8.4}$$

Image rejection is made easier if the image frequency can be kept as far away from the desired frequency as possible. This places it far down on the attenuation curve of the filter. This positioning means that high intermediate frequencies should be used. But, if they are too high, the same filter develops a problem on its lower edge—it can't keep strong signals from directly entering

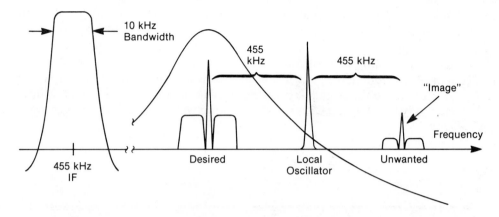

Figure 8.8 Frequency spectrum showing the relative positioning of the IF passband at 455 kHz, the local oscillator at 1.455 MHz, and the two signals that are located 455 kHz above and below the oscillator frequency. One of these will not be wanted and this is termed the *image*. It is eliminated by using a simple bandpass filter.

the IF. Images are normally more severe when the receiver is tuned to its highest frequency since the image is then "closer" on a percentage basis.

8.3 LOCAL OSCILLATOR

The variable frequency oscillator deserves some comment. Oscillators in general and variable-frequency ones were discussed in Chapter 5. For superhet applications, the oscillator is normally called the *local oscillator*, which simply means that it is inside the receiver. The requirements the oscillator must meet are fairly simple. It will be adjusted to be 455 kHz higher than whatever frequency is to be received and must stay there without drifting. If it does drift, the receiver will have wandered over to some other station.

For AM broadcast band work, it is normal to operate the local oscillator at a frequency higher than the desired signal. Therefore, to tune the full band from 540 to 1600 kHz, the oscillator must be tunable from 995 to 2055 kHz (assuming a 455 kHz intermediate frequency). The ratio of the highest to the lowest frequency for the oscillator is just over 2 to 1. This is a relatively easy design.

The oscillator could also be chosen to operate 455 kHz below the desired signal frequency. For the AM broadcast band, this would mean the oscillator would have to tune 85 to 1145 kHz. This is the same number of kHz from lowest to highest frequency that we had before but now the ratio of lowest to highest frequency is more than 13 to 1—a much harder design! This is the main reason the local oscillator is usually chosen to operate at the higher-frequency range.

Now for a legal problem. Since the receiver contains an oscillator, it could act as a small transmitter and, therefore, interfere with other radio receivers. All receiver designs should be checked to see how much of this signal radiates into the free space around it. Most countries have standards that place a limit on the strength of this radiated signal. These problems can be minimized by limiting the amount of DC power going to the oscillator circuit itself, using proper power supply filtering capacitors, reducing long wires and, finally, if necessary, shielding.

8.4 MIXER AND OSCILLATOR

Figure 8.9 gives an example of a mixer and oscillator for a superhet receiver. The mixer uses a field effect transistor as it can provide nearly ideal square-law mixing characteristics. At its gate terminal, a variable-tuned circuit selects the general range of RF signals to be received. (The exact band of frequencies is selected by the IF filters.) The main unwanted signal this tuned circuit must eliminate is the image frequency.

At its source, a strong oscillator signal causes the bias point to shift up and down, making the FET operate on different portions of its nonlinear characteristics. At its source terminal, a fixed 455 kHz filter is connected. The source current will contain many frequencies as a result of the mixing process. This filter will only pass the desired IF signals so they can undergo further amplification and filtering.

The oscillator shown is a tunable Colpitts design with feedback from its collector to the emitter. The output is taken from the emitter as this is a low impedance point in the circuit.

Both the oscillator and the input of the mixer circuit must be tunable—in fact, they must be tunable together, i.e., they must "track" each other and stay 455 kHz apart. The "ganged" tuning capacitor necessary to do this is also shown in Figure 8.9. The larger capacitance section will be used for the lower tuning frequencies at the mixer input and the small capacitance section will be used by the oscillator. Even with a special ganged capacitor as shown, the tracking adjustment can be difficult to design and set. Normally, a small set of screwdriver-adjust trimmer capacitors will also be included on the main tuning capacitor to ease the problem.

For reasons of economy, the two individual circuits can be combined into a single transistor circuit that oscillates and mixes at the same time. This is shown in Figure 8.10. The circuit is called a *converter* or *autodyne mixer*. It is a difficult design with which to obtain any degree of "linear" mixing and is normally used for inexpensive AM broadcast receivers.

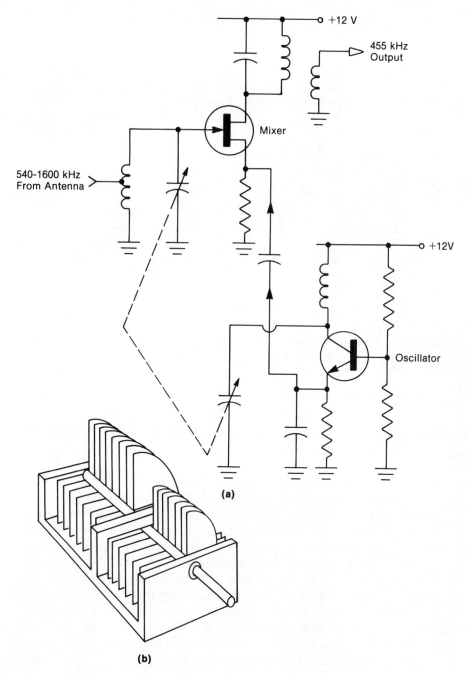

Figure 8.9 Schematic diagram (a) of the mixer and oscillator section from an AM broadcast receiver. The "ganged" tuning capacitor for the RF input and the oscillator is shown in (b).

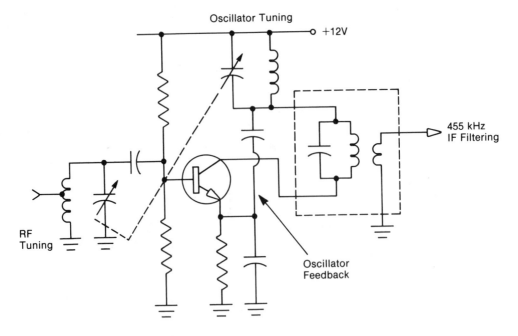

Figure 8.10 Autodyne mixer circuit combines the mixer and oscillator into one stage. Its main advantage is economy.

8.5 RADIO FREQUENCY AMPLIFIERS

The mixer section of a receiver does not have a very low noise figure. For receivers operating below about 3 MHz, this isn't a problem because there is so much noise external to the receiver that lower noise figures would be a waste of money. Broadcasters must compensate for this by using high transmitter powers and keeping their listeners fairly close to the antenna towers. However, above about 30 MHz, the external noise drops significantly and so receivers can benefit from improved noise figures and better sensitivity. Therefore, Citizen's Band, FM broadcast, and TV receivers, etc., normally include one or more radio frequency (RF) amplifiers ahead of the mixer. An example of this is shown in Figure 8.11. Notice the added complexity because now there are three tuned circuits that must track each other as the receiver is tuned—RF amplifier, mixer, and oscillator. However, the problem isn't as bad as it might appear because the RF and mixer-tuned circuits will always be tuned to the same frequency and so can use the same component values. The tuning capacitor will now need three ganged sections. The two larger capacitance sections will be the same.

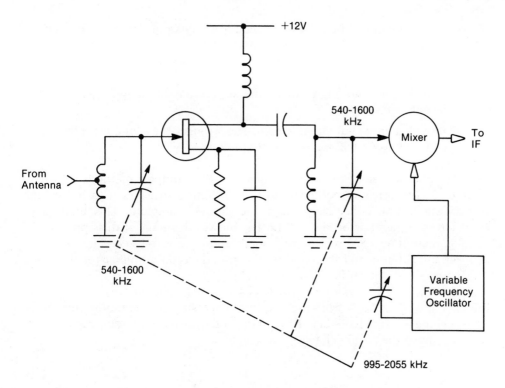

Figure 8.11 RF amplifier can be added ahead of the mixer to improve the receiver's noise figure. Now three tuned circuits must track. These circuits are collectively known as the receiver's *front end* or *tuner* section.

8.6 INTERMEDIATE FREQUENCY AMPLIFIERS

The intermediate frequency amplifiers provide the bulk of the receiver's gain and shape its passband response. This is the main section that determines the selectivity of the receiver. The front-end stages ahead of this are essentially the same for both an AM and an FM receiver. But, now the differences start to show up. We will look at an IF amplifier for an AM receiver first.

8.6.1 AM IF Amplifier

The front-end circuits of the receiver have done a preliminary selection of the approximate frequency range desired. This range has now been moved down to the intermediate frequency and the signals have been amplified somewhat.

However, there will still be a lot of unwanted signals present and the strength of the desired signal could be virtually anything, depending on the strength and distance of the transmitter. The AM IF must, therefore:

- Provide good filtering to shape the desired passband and reject any adjacent signals.
- Provide a large but variable gain. The stronger the signal from the front end, the less it will have to be amplified.

The filtering can be provided using any of the bandpass methods shown in Chapter 3. The original method is to use discrete transistors with individual tuned circuits that collectively shape the desired passband. This method is shown in the schematic of Figure 8.12. The big disadvantage of this approach is the alignment time needed. A more modern approach is to use ceramic, crystal, or mechanical filters and then add an integrated circuit amplifier for the gain. This approach is shown in Figure 8.13.

The discrete transistor IF shows *neutralizing capacitors* used to feed an out-of-phase signal back from the output of each stage to the input. This tends to cancel the internal feedback through the collector-base junction capacitance that can make the amplifier tend to oscillate.

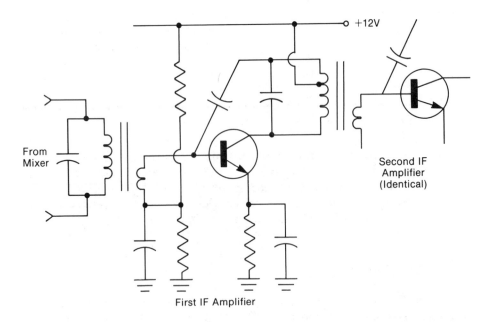

Figure 8.12 AM IF amplifier using discrete transistors and filters.

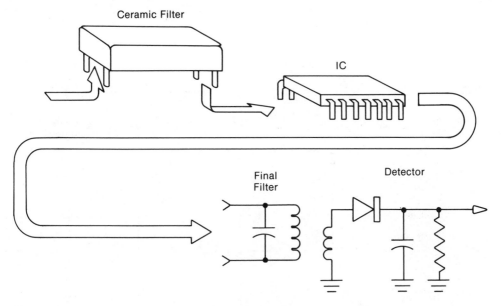

Figure 8.13   AM IF using a packaged filter with integrated circuit (IC).

8.7   AUTOMATIC GAIN CONTROL

As mentioned earlier, the signals entering the IF amplifiers could have a wide range of signal strengths and the gain must be adjusted accordingly. This is accomplished using an automatic gain control (AGC).

We saw back in Chapter 2, when we were talking about transistors, that the gain of a transistor depends on the bias current. Therefore, the gain of an amplifier can be made to vary if the bias point of the individual transistors can be adjusted. Since the purpose of an AGC circuit is to maintain a constant output level at the demodulator circuit, regardless of the incoming signal's strength, this is the point at which the reference voltage is picked up. With a demodulated AM signal, there is a DC component that is proportional to the carrier's strength and then there is the AC component that represents the modulating signal. For the AGC function, the DC component is used to reduce the gain of the IF amplifiers. The stronger the DC component at the demodulator, the more the gain is reduced. This is a feedback circuit so, as the gain is reduced, the DC component is also reduced—but not by as much. The DC component at the detector output will always be a bit larger for a strong IF input signal than it will be for a weaker one.

Figure 8.14 shows the operation of an AGC circuit in block diagram form. Here, the DC voltage at the detector output is filtered to remove the AC compo-

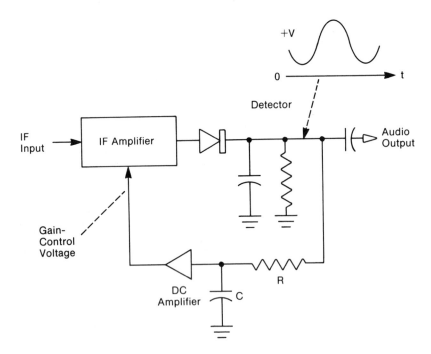

Figure 8.14 Automatic gain control (AGC) is used to reduce the gain of the receiver's IF amplifiers when strong signals are being received. This will keep the detector output more nearly constant.

nent and then amplified with the DC amplifier. This stronger voltage is now used to reduce the gain of the IF amplifiers. When there is no output at the detector, the IF gain should be at its maximum.

The resistor and capacitor used to filter the DC before amplifying deserve some comment. The RC time constant obviously must be large enough to remove the audio component or else gain will be fluctuating up and down in step with the lower audio frequencies. But the time constant must also accommodate the changing signal strengths that would be seen, for example, by a radio in a moving vehicle or a radio that is being tuned across the band and encountering different stations. An RC time constant of about one-half second is usually satisfactory.

Improved AGC operation can be obtained if two separate time constants are used—a short one when the signal strength is increasing and a longer one when it is dropping. This provides a very desirable "fast attack, slow decay" characteristic. SSB receivers normally must use such a dual time constant because there is no DC component at the detector output because there is no carrier coming through. Signal strength must, therefore, be guessed at by working with the audio strength.

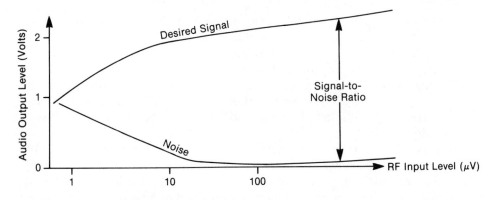

Figure 8.15   AGC operation will keep the total signal output fairly constant but the noise component will be reduced. As a result, the signal-to-noise ratio will improve.

As signal strength to the receiver increases, the signal-to-noise ratio at the detector output will improve. This is shown in Figure 8.15. The reason is that, although the total signal output will not be increasing by much, the amount of gain between the antenna terminals and the detector will have been reduced. Any unwanted noise either from external or internal sources will not be amplified as much.

A final question. In order to have a variable gain amplifier, it must be nonlinear. The change in gain is obtained, as we have already said, by moving to different slopes on this curved characteristic through a bias adjustment. But, if the input/output characteristic is curved, all signals, no matter how small, will generate harmonics and intermixing products—in other words—distortion. How is it possible to have variable gain without any distortion? The answer is found by looking at the different frequencies that the sums, differences, and harmonics would appear at. These would all be at audio and at two and three times the IF frequency. As long as some bandpass filtering at the IF frequency exists after the AGC stages and before the detector, these products will never appear. IF frequencies must be selected high enough so that the upper side of the passband is much less than double the frequency of the lower side. Therefore, filtering out of the harmonics will be easy.

8.8   FM INTERMEDIATE FREQUENCY AMPLIFIERS

If the receiver is FM, several differences show up in the IF section. First, the bandwidth must be wider, which simply means a change in the filters. If a higher IF frequency is used, the wider bandwidth is easy because the loaded Qs of the coils will remain nearly the same. Second, the filters must have a more constant phase delay across the passband. This does make the filters significantly harder to design. But the big difference for an FM receiver is that the amplifiers now are "limiting amplifiers." They are given extremely high gains so they intentionally go into cutoff or saturation. This removes any amplitude variation that the signal may have acquired in its travels and leaves only the frequency variations. These are demodulated by some type of FM demodulator (see Chapter 6).

Virtually all FM IF amplifiers now use an integrated amplifier. An example of a typical circuit is shown in Figure 8.16. The emitter-coupled pairs form differential amplifiers that are biased at relatively low current levels. With symmetrical signal inputs, the individual transistors will go into cutoff long before they will go into saturation. Symmetrical limiting takes place by alternately cutting off one transistor in the pair and then the other. For very weak signals, the first two pairs will act as linear amplifiers and the last pair will perform the limiting. As the input signal strength increases, the limiting will occur in stages closer and closer to the input.

No AGC is needed for these amplifiers as overloading is now a desirable feature and the limiting action keeps the detector output constant regardless of signal strength. AGC may still be needed for the RF amplifier to keep it from overloading either itself or the mixer circuit. AGC voltages cannot be obtained from the detector in an FM receiver because all amplitude variations will have been removed by the limiters. Instead, most integrated circuits will provide a signal strength output that has been derived from the average current through the limiter amplifiers.

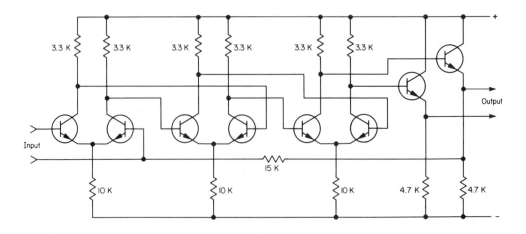

Figure 8.16 Limiter amplifier used for FM IF amplifier. The circuit is available as an IC.

8.9 FM CAPTURE EFFECT

What happens if two FM signals happen to be on the same frequency and one is just a bit stronger than the other? This could happen if you live about half way between two distant cities each with a broadcast station on the same frequency. The answer is that only the stronger station will be heard and the other will be completely blocked out—provided a minimum difference exists between the two signal strengths. This minimum depends on the individual receiver and on its amplifiers in particular.

Even though the two transmitters are on the same assigned frequency, their carriers will often be on slightly different frequencies at any instant because their modulation will be different. The two signals will, therefore, perform a vector addition with a result that looks like an AM signal. But, since one signal is stronger than the other, there will never be a complete cancellation. The limiter amplifiers will remove this amplitude variation leaving only the stronger station. This also explains why the sidebands from an adjacent channel can be allowed to extend into the one you want to listen to. A good FM receiver will have a capture ratio of about 1.0 dB. This means that, as long as the one station is at least 1.0 dB stronger than the interfering station, it will be heard.

8.10 AUTOMATIC FREQUENCY CONTROL

At the higher frequencies that FM broadcast tuners and UHF television operate at, the drift of the local oscillator becomes a problem. The receiver will require constant fiddling to keep it on the desired station. One method that can be used to prevent this is to use a quartz crystal for the local oscillator. However, this makes changing stations very difficult. A variation on this is to use a phase-locked loop synthesizer that is based on a quartz crystal reference. We will examine this in the next chapter.

A fairly simple solution is to create some sort of feedback system that will move the oscillator back on frequency when it wanders. But such a system needs a reference. This can be obtained by checking to see that the signal is centered in the middle of the IF passband. Regardless of whether the receiver is for FM or AM operation, a discriminator (Figure 6.30) can be used to generate a DC voltage proportional to the average frequency of the signal within the passband. This voltage can then be used to alter the capacitance of a varactor diode and this, in turn, will shift the oscillator back, almost, to where it should be. (If it were shifted exactly back, there wouldn't be any DC voltage left to do the shifting.) The system is more stable than the oscillator alone since the tuned circuits in the discriminator will be operating at the IF frequency which is

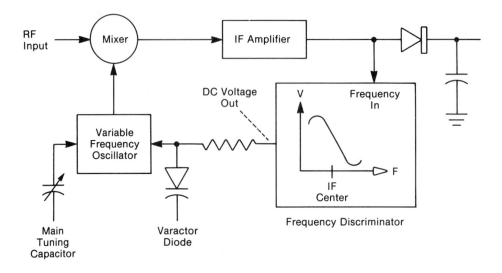

Figure 8.17 Automatic frequency control (AFC) can be added to any VHF or UHF receiver. An error signal from an FM discriminator alters the capacitance of a varactor diode to correct the drift.

much lower than the frequency the tuned circuit in the oscillator is operating at.

8.11 RECEIVER TESTING

As we did with the previous chapter on transmitters, the last section of this chapter describes some of the tests that should be performed when a new receiver design is being worked on or an existing receiver is being repaired. Most of the characteristics to be tested have already been defined.

8.11.1 Selectivity

Selectivity is a test of the receiver's passband characteristics and is mainly determined by the IF filters. The front-end circuits of the receiver play very little part in this. The test is simple to perform if a good sweep generator is available. It is connected to the antenna terminals and the display unit is connected to the demodulator output. However, there could be a few problems. First, if this is an AM receiver, the AGC will have to be defeated. If it isn't then, as the swept signal gets weaker near the very edges of the passband, the AGC will allow the gain to increase and so produce inaccurate results. Another possible problem with an AM receiver is its demodulator; it may not produce an output that is proportional to its RF input. In this case, a calibrated attenuator will have to be used initially to set up reference marks on the display device.

The problems increase for FM receivers. The signal cannot be passed through the limiter amplifiers because they would simply chop off any amplitude variations and make the response appear to be very broad and exceptionally flat. If the filtering is all located in the one package ahead of any limiters, the signal can be detected at that point with a separate RF demodulator probe.

8.11.2 Sensitivity

This is a test of how weak a signal the receiver can pick up without having it obliterated with noise. It will depend on the noise figure of the receiver's front end and also the bandwidth of the IF filters. Therefore, any major repair work on either of these sections should be followed by a sensitivity check.

The test essentially consists of connecting a modulated signal generator to the receiver's input and then reducing the signal strength until the modulation just starts to fade into the noise. The generator output in microvolts is then the receiver's sensitivity. Actually, the test should be done more carefully than this because it isn't very scientific to say exactly when "the output just starts to fade into the noise." Often, then, an audio meter is connected to some convenient point in the audio circuit of the receiver. With no RF signal applied, the audio noise level is noted. Then the RF signal, with 50% modulation, is increased in level until the audio increases by some specific amount (usually 10 dB). A typical sensitivity might then be 2.3 microvolts of signal for an output signal-to-noise ratio of 10 dB. A more precise definition of *sensitivity* is found in the Electronic Industries Association's (EIA's) Standard RS-204A. The modulated RF signal level is increased until the ratio of the total audio output (modulation plus noise plus audio distortion products) is 12 dB stronger than the noise plus distortion products alone. Sensitivity measured this way is called the *12 dB SINAD ratio* (SIgnal, Noise, And Distortion). It includes the inevitable problem that some distortion of the modulated signal will occur some place in the receiver and this distortion will just be as bad as random noise. The test equipment is a bit more complicated since the one reading must measure the total audio output with the desired modulation component removed using a sharp notch filter. Several manufacturers make a SINAD meter that includes all of the necessary amplifiers and filters.

8.11.3 Spurious Signals

The whole idea of the receiver is to pick up a small desired range of frequencies and get rid of anything else. However, there may be a few other frequencies

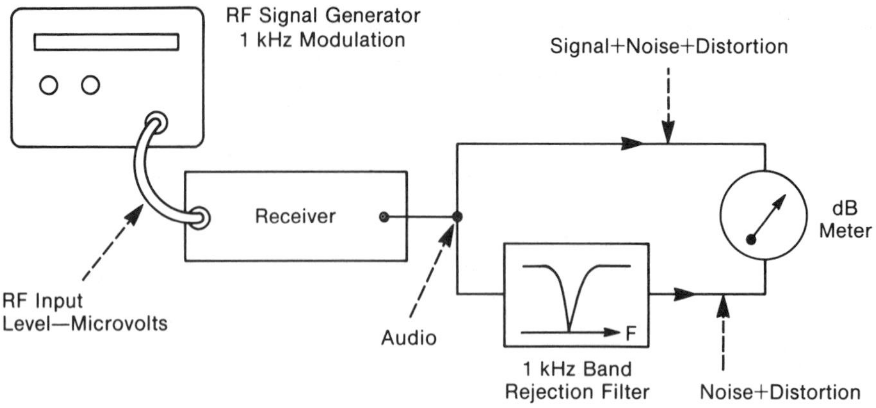

Figure 8.18 SINAD test for receiver sensitivity involves the measurement of two audio output levels—one with full audio output and the other with the modulation tone filtered out but leaving any harmonics that may be the result of distortion in the receiver.

that could sneak in for various reasons. The simplest check is to set the receiver to one frequency. Connect a variable frequency RF generator, with its modulation turned on, to the receiver's input then tune the generator across a wide range of frequencies and listen for unwanted outputs. Prime frequencies to check are the **IF** frequency itself and the image frequency.

A more elaborate check requires the use of two signal generators to check for cross-modulation. Both generators are connected, through a resistor mixing circuit, to the receiver's input terminals. One generator is set, without modulation, to the same frequency that the receiver is tuned to. The other one, with full modulation and a strong RF level, is tuned over a wide range above and below the desired frequency. Any stray signals received this way will be the result of accidental mixing in the very early stages of the receiver.

8.12 REVIEW QUESTIONS

1. Describe sensitivity. Which portions of a receiver are most involved in determining its sensitivity?

2. Describe selectivity. Which portions of a receiver determine this characteristic?

3. Why is good sensitivity of little importance for radio signals below about 3 MHz?

4. What are the major defects of a "crystal set"?

5. What problems of the TRF receiver are solved with the superhet receiver?

6. A crystal set consists of a single tuned circuit. The inductance is 200 μH and the capacitor is variable from 400 down to 60 pF. The resistive load on the tuned circuit due to antenna, diode, and earphones is a constant 10,000 ohms. What is the resonant frequency and bandwidth at the two extremes of the variable capacitor?

7. In a superheterodyne receiver, the local oscillator is set to 1.35 MHz. The intermediate frequency amplifiers are tuned to 455 kHz. What three possible frequencies could enter the mixer and pass into the IF amplifiers?

8. A superhet radio receiver picks up a certain station when its dial is set to 11.9 MHz and again, but weaker, when the dial shows 10.1 MHz. Why does this happen? What is the true frequency of the station? What is the receiver's IF frequency?

9. What are the image frequency and IF frequency for a receiver tuned to 5.2 MHz and with a local oscillator running at 4.15 MHz?

10. Give three factors that must be considered when selecting the IF frequency for a new receiver.

11. What would be needed if AFC were to be added to an AM receiver?

12. Describe the operation and the benefits of AVC when applied to an AM receiver.

13. What is *tracking*?

14. Which portion of an FM receiver determines its capture ratio?

9. PHASE-LOCKED LOOPS

The phase-locked loop (PLL) has become an essential building block in many different areas of electronic communications. With very little variation, it can act either as a stable oscillator that can be adjusted in small frequency steps or as a narrow bandpass filter that can "track" a changing input frequency.

Much of the circuitry used for electronic communications has remained essentially unchanged over the past 50 years. It is true that a lot of modernizing occurred with the introduction of the transistor and the integrated circuit but the basic circuitry principles really didn't change all that much. The phase-locked loop is the real advancement that caused significant changes and improvements. In its integrated circuit form, this sophisticated circuit has become a low-cost, high-performance, but easy to use building block.

9.1 PLL APPLICATIONS

To illustrate the versatility of the PLL circuit, consider the following list of applications:

1. As the 15,734 Hz (North America) synchronized horizontal scanning oscillator in all TV receivers.

2. As the 3.579 MHz (North America) synchronized color oscillator in all color TV receivers.

3. For FM demodulators in the sound section of TV receivers and for FM broadcast receivers.

4. As part of the channel demodulation circuitry in FM stereo receivers.

5. For a video detector (AM) in some high-quality TV receivers.

6. As a frequency "synthesizer" to replace the local oscillator in TV, CB, AM, and FM broadcast and aircraft NAV-COM receivers. This application is usually coupled with a digital frequency display for the "high-tech" look.

7. As selective tone filters in remote control, telephone, and data communication circuits.

By now you should have the general idea. This is an important device.

9.2 THE BASIC PHASE-LOCKED LOOP

The basic loop is shown in block diagram form in Figure 9.1. This shows the three main components that are present in all applications. For some applications, this is all that is needed. For others, like frequency synthesis, extra circuitry must be added. The diagram is especially fitting because the two blocks are generally available as integrated circuits—either individually or combined—and the third component, the filter, is often built with discrete components to suit the specific job requirements.

The operation is very easy to summarize. The "frequency output" will be the same frequency as the "input" but the output amplitude will be fixed and there may be some phase shift. The "DC output" will be some constant voltage that is proportional to the output frequency.

So far this doesn't sound too exciting. It sounds like the basic operation is roughly the same as a piece of wire stretched between input and output. After all, if the input frequency changes, the output will change to the same value. It doesn't look as if we have accomplished very much. However, the list of applications seemed to suggest that the circuit is quite useful so we had better dig a

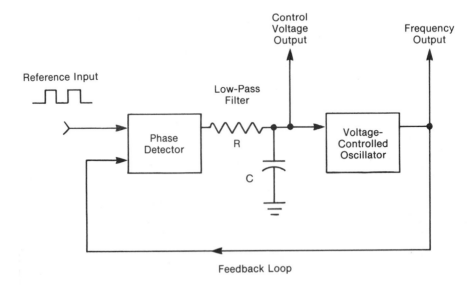

Figure 9.1 Three essential components of the basic phase-locked loop. The circuit has one input and two possible outputs. Both the phase detector and the VCO are available in low-cost integrated circuit form.

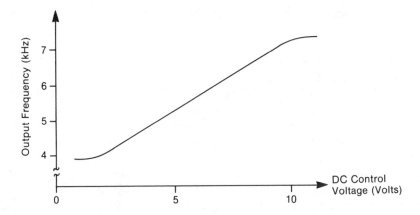

Figure 9.2   Sample relationship between the DC control voltage at the input of a VCO and its output frequency.

little further. Let's look quickly at the three basic components and then see how they work toegther.

One of the components is the *voltage-controlled oscillator* (VCO). It is an oscillator whose output frequency is adjusted by a DC control voltage applied to its input rather than through a manual adjustment. We first looked at VCOs back in Chapter 5. They will be examined in more detail later in this chapter. For PLL applications, the relationship between the control voltage and the output frequency is important. Such a characteristic is shown in Figure 9.2.

For some applications, the sine wave or square wave produced (depending on the individual circuit) by the VCO will be used as the PLL output. This is used when the loop is acting as an oscillator or a filter. For other applications, the DC control input to the VCO is used as the output of the PLL circuit. FM demodulator circuits are typical of this configuration.

Regardless of the application, part of the oscillator signal is fed back to the *phase detector*. This is the "loop" portion of the PLL's name. The phase detector compares the VCO signal with the "reference input" signal. When the loop is operating properly, the two inputs will be at the same frequency and the loop will be "locked." Under this condition, the phase detector will produce an output whose average DC voltage will be proportional to the phase difference between the two inputs. What happens when the two are not the same frequency is a bit more complex and depends on the actual circuitry used for the phase detector. We will examine these variations a little later.

The phase detector output is normally a series of pulses of constant amplitude. The ratio of Hi to Lo time will change as the relative phase of the two input signals changes. These pulses must be filtered to obtain the average DC component. The RC low-pass filter serves this purpose. Typically, it is no more elaborate than this.

In overall operation, the VCO in the PLL circuit will always try to run at the same frequency as the input signal. This is called *tracking*. If the input changes frequency slowly, the VCO will follow. This happens because the phase detector will be able to sense that the reference has altered its frequency but the VCO signal will not immediately change. Therefore, there will be a momentary phase difference and this will result in a change in the DC control voltage. This voltage will move the VCO (hopefully) in the right direction and so the two signals will again be identical. Figure 9.3 shows that the control voltage will move up and down in step with the changing input signal. The control voltage actually represents the demodulated version of the changing input frequency. We now see how the PLL can be used as an FM demodulator.

If the input frequency moves too far from its initial value or too fast, the VCO will not be able to follow. The loop, in this case, will lose phase lock. But the circuit will attempt to regain the lock condition. In the process, the control voltage may go through a short period of oscillations as shown in Figure 9.4.

The low-pass filter plays an important part in the operation of the loop as it helps to determine the dynamic characteristics of the circuit. The filter determines how fast the circuit will re-establish lock, how stable the lock will be and, in some cases, whether lock will even occur. A full analysis would involve considerable mathematics and a thorough understanding of the principles of control systems and feedback loops. In this chapter we are only after a basic understanding.

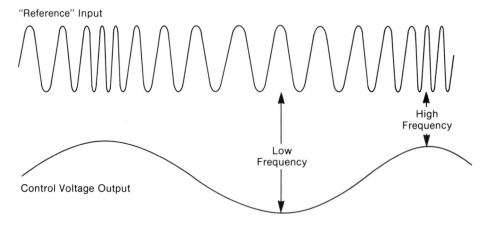

Figure 9.3   Changing input frequency will result in a changing control voltage. The result is an FM demodulator.

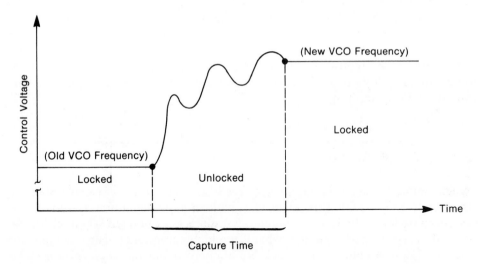

Figure 9.4 The PLL circuit may lose phase lock if the input frequency changes too far or too fast. Then the control voltage will go through a period of oscillation until lock is re-established.

9.3 TRACKING RANGE

There is a limit to how far the input frequency can move before the loop stops tracking and stays out of lock. The highest and lowest frequency to which the VCO will follow the input (when it is changing slowly) will determine the *tracking range*.

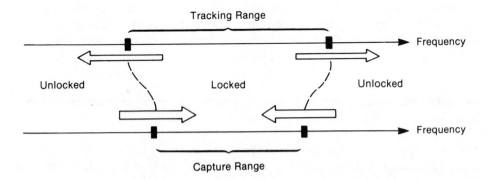

Figure 9.5 Comparison of tracking range and capture range of a phase-locked loop. Tracking range is measured with the loop initially locked. Capture range is measured with the loop initially unlocked.

To measure this, the input signal is set to some middle frequency where the loop is initially locked. Then the input is moved slowly to each extreme until lock is lost; a dual-channel scope will show that the input and output are no longer the same frequency. One low-frequency and one high-frequency limit define the tracking range. The VCO circuit itself is mainly responsible for determining the tracking range.

9.4 CAPTURE RANGE

When an input signal is first applied to the PLL circuit, it may or may not cause the loop to lock. To find this range, start with an input signal that is initially too low to cause lock. Then slowly increase it until lock is obtained. Then repeat the procedure, starting with an input that is too high a frequency. The one high and the one low frequency where lock is obtained define the *capture range*—the range where the VCO will be "pulled over" to the input frequency.

The capture range is normally narrower than the tracking range. The two are summarized in Figure 9.5. Notice the difference in the directions the input signal must be moved when the measurements are made. All three components collectively determine the capture range.

9.5 CAPTURE TIME

When an input signal within the capture range is suddenly applied to the PLL, the circuit will take a certain amount of time to acquire phase lock. This capture time is the reason that both capture range and tracking range had to be measured with the input frequency changing slowly. If the input had moved too fast, the limits would have been exceeded long before the loop showed that lock had been lost. During this capture period, the control voltage may go through a period of oscillation as shown in Figure 9.4. Capture time is dependent on all components in the loop but mainly on the filter time constant.

9.6 PLL AS A FILTER

The only application that we have described so far is the use of the phase-locked loop as an FM demodulator. In that case, the output was taken from the DC control voltage point. What do we get if we use the VCO output?

It has already been suggested that the PLL in this configuration looks like a piece of wire. If the input is 800 kHz, the output will also be 800 kHz. If the input changes frequency, the output will likely follow.

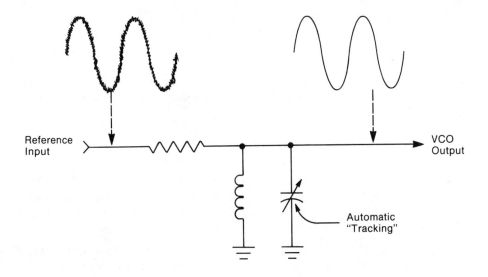

Figure 9.6 The PLL can operate much the same as a very high Q bandpass filter. The filter automatically adjusts its center frequency to "track" the input frequency.

The important point is that the VCO output frequency is the "average" of the reference input frequency and it is the low-pass filter that does the averaging. The lower the cutoff frequency of this filter, the longer the averaging time. The reference input could, therefore, be a bit noisy—both in amplitude and frequency jitter—and the output will not be affected. The PLL is, therefore, operating as a very high Q bandpass filter. The two simple components, the resistor and the capacitor, determine the bandwidth and this is quite independent of the operating frequency.

There are some differences, however, between the PLL as a bandpass filter and a more conventional tuned circuit. One is that the output can only be a single frequency and so it cannot be used to pass a "band" of frequencies. The output also cannot change amplitude and so any input amplitude variations are ignored. Finally, the filter moves. Since the VCO will track the slow changes in the input frequency (but not the fast changes), the "filter" will tune itself to follow the input.

9.7 VOLTAGE-CONTROLLED OSCILLATORS

Now we are going to take a more detailed look at the circuitry inside the various blocks, starting with the voltage-controlled oscillators.

One circuit was looked at in Chapter 4. This was a varactor-diode-controlled circuit suitable for a very limited frequency range. Now we will look at some other circuits—ones more suited to integrated circuit form and capable of being controlled over a wider frequency range.

The essential characteristics of any VCO are summarized in Figure 9.2. When the equivalent measurements are made for another VCO, the resulting graph will show:

- The highest and lowest frequency between which the oscillator can be controlled. This will determine the loop's tracking range.

- The "linearity" of the control curve. This is very important if the VCO is to form part of an FM demodulator because any departure from a straight line here will result in distortion during the demodulation process.

- The "gain" of the VCO. This is a number that relates the output change in Hertz to the input change in volts. This number is important when the mathematics of loop stability is being played with.

A VCO based on a two-transistor multivibrator is shown in Figure 9.7. This is a conventional multivibrator[1] circuit except that the base bias resistors go to a separate, variable control voltage instead of to the same power supply as the collector resistors. The variable bias supply will be able to control the off time of each transistor. A high-control voltage makes the capacitors charge faster than a lower control voltage so the transistors will turn back on sooner. The result will be a higher frequency.

The input-output characteristics of this circuit are shown in Figure 9.8. Over a wide range of control voltages, the output frequency changes linearly. For very low voltages, the circuit just stops oscillating because there isn't enough bias current to turn the transistors on. For high voltages, the reverse problem occurs; there is too much base current and the transistors will not come out of saturation.

For control voltages between 3 and 7 volts, what is the "gain" of this circuit?

---

[1]In the multivibrator, only one transistor is on at a time— in full saturation. The other transistor is cut off. When the changeover takes place, the collector of the transistor that has just saturated will have dropped from the full supply voltage down to zero. The base of the other transistor will be at the opposite end of the coupling capacitor so its voltage must change by the same amount—from about +0.6 volts down to a negative value roughly the same magnitude as the collector supply voltage. Now the base voltage slowly rises as the capacitor charges toward the control voltage. But the transistor turns back on when its base reaches +0.6 volts and the cycle repeats.

The charging time for the VCO changes because the starting point is always the same (the negative equivalent of the collector supply) but the voltage it is trying to charge to is variable (the control voltage).

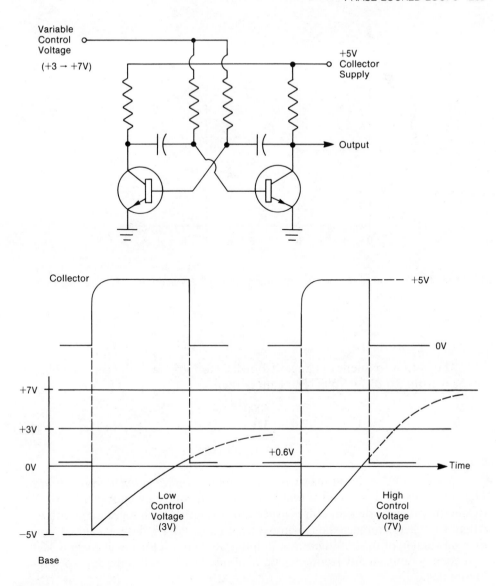

Figure 9.7 Voltage-controlled oscillator based on a multivibrator. The collector and base waveforms for one transistor are shown for two different values of control voltages.

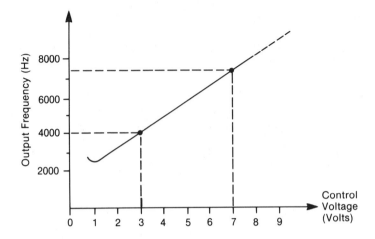

Figure 9.8   Variation of frequency with changing control voltage for the multivibrator in Figure 9.7.

The output frequency changes from 4000 to 7500 Hz as the control voltage changes from 3.0 to 7.0 volts. The gain is then:

$$= \frac{7500 - 4000}{7.0 - 3.0}$$

$$= 875 \text{ Hz per volt (Hz/v)}$$

A second VCO circuit uses what is called an *emitter-coupled multivibrator*. The advantage of this circuit is that it requires only one capacitor instead of the two in the previous design. This is especially attractive for integrated circuit use where the users connect their own external capacitor to the package. A commercial version of this multivibrator in integrated circuit form will cover a 3.5 to 1 frequency range at any frequency up to about 30 MHz. The circuit is shown in Figure 9.9.

The multivibrator's operation can be briefly described as follows. The two lower transistors act as variable current "sinks" with the actual amount of current set by the control voltage at their bases. As current sinks, they can control the charging rate of the capacitor above them. The upper two transistors act something like a flip-flop with only one transistor on at a time. One end of the capacitor will be held high by whichever transistor above it happens to be on. The other end will charge down toward ground. At some point, it will have reached a low enough voltage to turn the transistor above it on and then the cycle repeats.

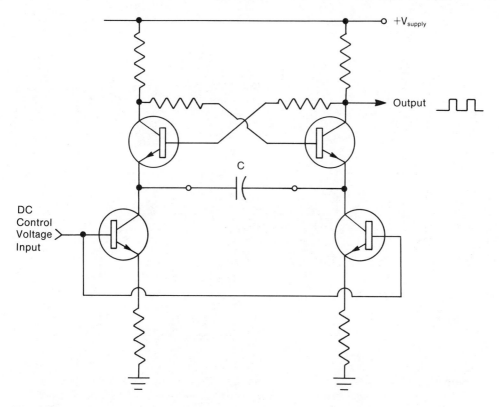

Figure 9.9 Voltage-controlled oscillator using an emitter-coupled multivibrator. Only one capacitor is needed. Control over a 3.5 to 1 frequency range is possible.

9.8 PHASE DETECTORS

Phase detectors are available in an even greater variety than the VCOs and actually fall into two different groups. Which group depends on what the detector's output does if the two inputs to the detector are not the same frequency. We will look at a number of candidates for the simplest group first and then a single, elaborate circuit for the second group. We will look at the circuits first and then answer the question of the two different frequencies later. All circuits are available in integrated circuit form.

9.8.1 The AND Gate

All of the basic logic gates can be used as phase detectors although some work better than others. Consider the operation of one of the poorest—the AND gate.

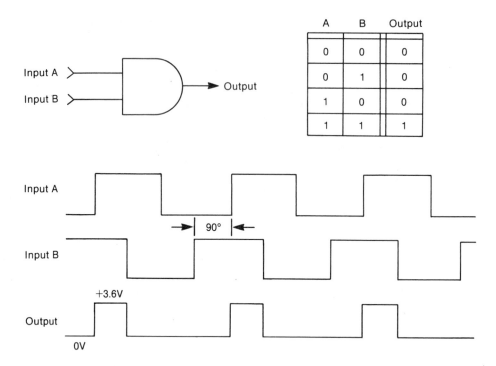

A	B	Output
0	0	0
0	1	0
1	0	0
1	1	1

Figure 9.10 The AND gate as a phase detector. With the help of the truth table, the output waveform can be drawn for the case where the two inputs are 90° out of phase.

As shown in Figure 9.10, each input has a square wave applied to it (Hi for 50% of the time and Lo for the remaining 50%). The relative phase angle between them could be anything. In this case, it is shown as 90°. The truth table can be used to draw the output waveform for any combination of inputs. It shows that the output will only be Hi when both inputs are high at the same time.

Assuming that a standard TTL (transistor transistor logic), gate is used, the normal Hi output is about +3.6 volts. For a 90° phase difference between the two inputs, the output will be Hi for one quarter of the time. The "average" output (after the low-pass filter) will be:

$$1/4 \text{ of } 3.6 \text{ volts} = 0.9 \text{ volts}$$

Now, instead, if both inputs were exactly in phase, the output would be Hi half of the time. The average output for this case would be:

$$1/2 \text{ of } 3.6 \text{ volts} = 1.8 \text{ volts}$$

Similarly, for inputs exactly 180° out of phase, the average output would be 0 volts.

Using these values, a graph can be drawn to show how the average output voltage would vary as the two inputs changed their phase.

Remember that although we keep talking about the average output voltage, the actual detector's output is a series of pulses at the same frequency as the input signal. The averaging and filtering will be provided later by the low-pass filter.

As with the VCO circuits, there is also a "gain" value for the phase detector. This relates the change in the average output voltage with a corresponding change in the input phase. For the AND gate this gain is calculated as follows:

For an input phase change from 0 to +180°, the average output changes from 0 to +1.8 volts.

$$\text{Phase detector gain} = \frac{\text{Output Change}}{\text{Input Change}}$$

$$= \frac{1.8 - 0}{180 - 0}$$

$$= 10 \text{ millivolts/degree}$$

If the same diagram is drawn for the logic NOR gate, it will be found to operate exactly the same—as far as phase detector operation is concerned.

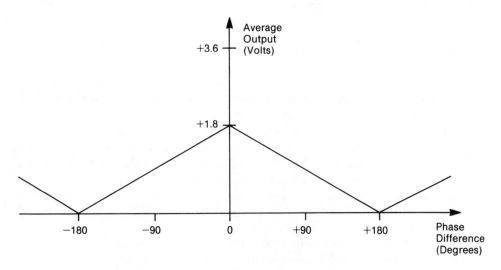

Figure 9.11 Average output voltage from the AND gate as the input square waves change phase between −180 and +180°.

Now for the disadvantages of the AND and NOR circuit:

1. The gain is positive between $-180°$ and 0 and it is negative for the other 180°. Depending on the rest of the connections in the PLL circuit, this will make the loop alternate between negative feedback (a stable condition) and positive feedback (a potentially unstable condition). The loop will still work but, every time a positive feedback situation occurs, the VCO will be pushed to a new frequency and phase angle to restore the negative feedback situation. It just makes the loop that much harder to design properly.

2. The gain is a bit low, which may or may not be a problem, depending on the other gains in the total loop. Other detectors have higher gains.

3. The output pulse is the same rate as the input signals. For some low-frequency applications, this will make filtering difficult.

4. Both inputs have to be a 50% duty cycle or the characteristics of the detector will change.

9.8.2 OR Gate Phase Detector

The OR gate is no better as a phase detector, and the NAND gate operates in a similar manner. The operation of the OR gate is shown in Figure 9.12. The only real change is that the average output voltages will lie between 1.8 and 3.6 volts if TTL gates are used. This is no immediate advantage but could suit the biasing requirements of some particular circuits.

9.8.3 Exclusive-OR Gate Phase Detector

This gate offers some significant improvements over the previous four gates when used as a phase detector. It is frequently used in many of the simpler PLL circuits. Its operation is summarized in Figure 9.13.

Two advantages can be spotted when looking at these characteristics. First, the overall gain is higher because the average output moves between a full logic Hi and Lo voltage. The second advantage is that the output pulse rate is double whatever the input pulses are. This may make the filtering and averaging easier. There are still two disadvantages left—the requirement for a 50% duty cycle on the input waveforms and the existence of both a positive and negative gain in the overall characteristic.

Actually, this circuit should look fairly familiar by now. It was introduced in Chapter 5 as a double-balanced modulator and then in Chapter 6 as a phase detector for an FM demodulator (a non-PLL version). Both of these applications used the "analog" equivalent of the gate which will accommodate lower

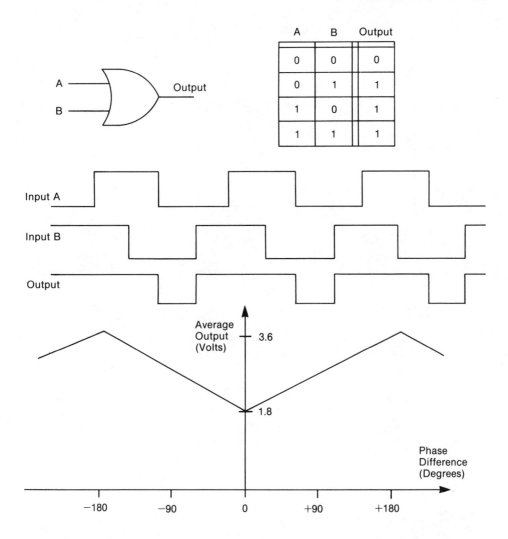

Figure 9.12 The logic OR gate as a phase detector. The average output voltages all lie between one-half and the full logic Hi voltage.

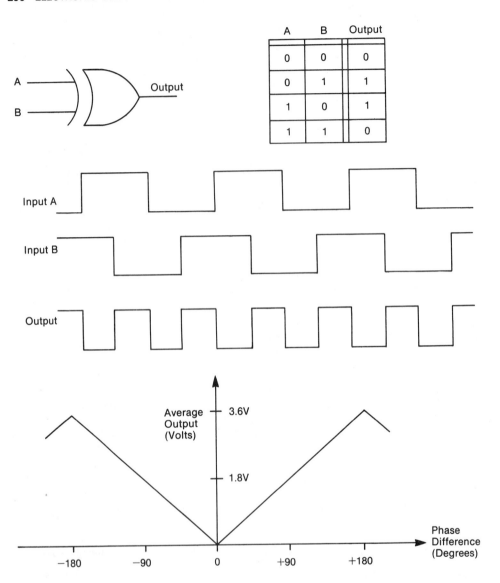

Figure 9.13 Exclusive-OR gate as a phase detector provides twice the output voltage and an output pulse rate that is double the input.

amplitude sine waves and also makes the output amplitude dependent on the input amplitudes. Most PLL applications use the "digital" version.

9.8.4 R-S Flip-Flop Phase Detector

The last circuit to be described within the first group of phase detectors consists of a RESET-SET (R-S) flip-flop. The Q output terminal will go Hi and stay Hi when a short (in time) pulse is applied to the SET terminal and the Q terminal will go Lo and stay there when a short pulse is applied to the RESET terminal. The operation is summarized by the three sets of input and output waveforms and by the average output voltage versus phase graph shown in Figure 9.14. The biggest improvement is the extended range of phase angles and the lack of

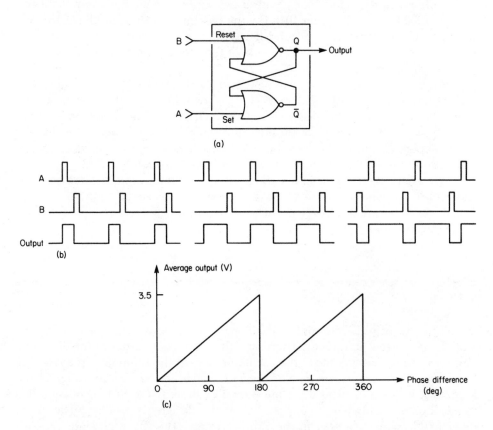

Figure 9.14 Phase comparator consisting of a SET-RESET flip-flop. The logic equivalent is shown in (a), sample inputs and output waveforms are shown in (b), and the circuit operation is summarized in (c).

any negative gain. The disadvantage is the necessity of generating the narrow pulses and the fact that this detector is more noise-sensitive because the narrow pulses are indistinguishable from noise.

9.9 DIFFERENT INPUT FREQUENCIES

All the detectors discussed so far belong in a group that has almost the same answer to the question—what happens if the two input signals do not have the same frequency?

If the two inputs have different frequencies, their relative phase angle will be continually changing over the full 360°. This will be happening at a rate that is the difference frequency between the two input signals. For example, if one input is 1000 Hz and the other 1010 Hz, the two inputs will be in phase, for very brief periods, 10 times every second. Then they will slowly drift out of phase by increasing amounts until, a short time later (one-tenth of a second), they will be back in phase again.

After filtering, the output will be the 10 Hz sine wave. The phase detectors are, therefore, also acting as mixers. This is shown in Figure 9.15 for the Exclusive-OR gate. All the other basic gates (AND, OR, NAND, and NOR) also operate the same although their output amplitude will be lower. The flip-flop version operates almost the same but the 10 Hz "sine wave" will have an abrupt discontinuity in it every time the two inputs pass through 0°.

We are going to pause in our description of different phase detectors for a moment to see what effect this first group has on PLL circuit operation. There is still one elaborate detector to be described and it has a rather different effect on the overall loop.

9.10 INITIAL CAPTURE

Each time the full PLL circuit is turned on, it would be very unlikely to find the VCO at the correct frequency. Therefore, the loop has to attempt to generate a control voltage that will move the oscillator toward the reference frequency so that phase lock occurs. As long as the VCO is within the capture range of the loop, lock will occur—after a certain period of time. If the VCO is outside this range, lock probably won't happen. This is the designer's problem. The phase detector has a part to play in the capture range. The five logic gate versions operate about the same. The flip-flop version will double the capture range.

To show how capture occurs, consider first the situation where the reference and the VCO frequency are very close together. The control voltage to the

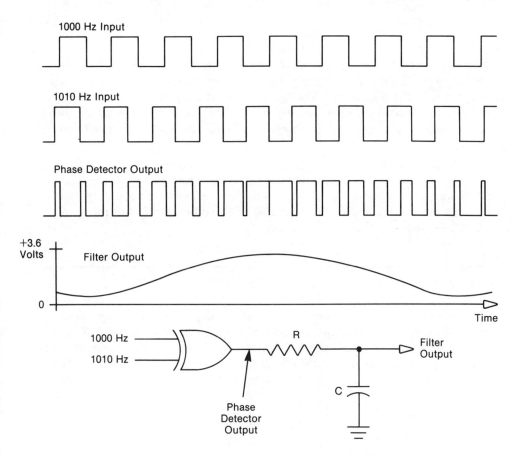

Figure 9.15 When the two inputs to a simple phase detector have different frequencies, the output (after filtering) will show the difference frequency. An Exclusive-OR gate is shown but the same is also true for AND, OR, NAND, and NOR gates.

VCO would contain a large amplitude difference frequency as long as this was within the passband of the low-pass filter. The VCO would be quickly moved to the correct frequency within one-half cycle of the difference frequency.

If the reference and VCO frequency are much further apart, but still within the capture range, a very high difference frequency would exist—so high it would be beyond the passband of the low-pass filter. But the filter doesn't cut off abruptly and so a very small amplitude signal would still pass through. The signal would be too small to cause immediate lockup but it would cause the oscillator frequency to move, alternately toward and then away from the correct frequency. But this doesn't look as if lock would ever be achieved. However, these small changes in VCO frequency are significant. It means that the relative phase shift will be moving a bit more slowly when the VCO is moving in the right direction and more rapidly when it is going the wrong way. The typical transient waveform seen on the control voltage line would look like that shown in Figure 9.16. The unequal shape of the peaks will produce an average component that will slowly push the VCO to the correct frequency.

Although many peaks are shown, remember that this is now a high-difference frequency so capture time should not be significantly different from the previous case.

Two final comments on this group of phase detectors. First, when the loop is locked there will always be some phase shift between the two inputs and this will change with frequency. Second, and more serious, is the possibility that the loop may lock with one input a harmonic of the other. If the full range of the VCO is less than 2:1, this won't happen.

9.11 FREQUENCY AND PHASE DETECTOR

All of the previous circuits operated as phase detectors when the two input signals were at the same frequency and as mixers when they weren't. This next circuit also acts as a phase detector when the two inputs are the same but when they aren't, the operation is different. If the one input is at a lower frequency than the other, the control voltage output will sit at its maximum instead of

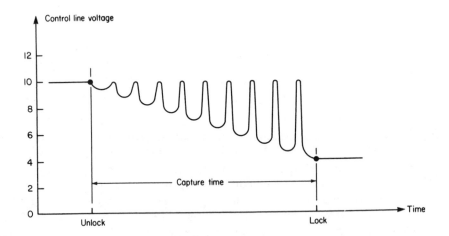

Figure 9.16 Transient waveform generated on the VCO control line as the loop pulls the VCO toward the correct frequency. This is common to all of the simple phase detectors. The starting point must be within the capture range of the loop.

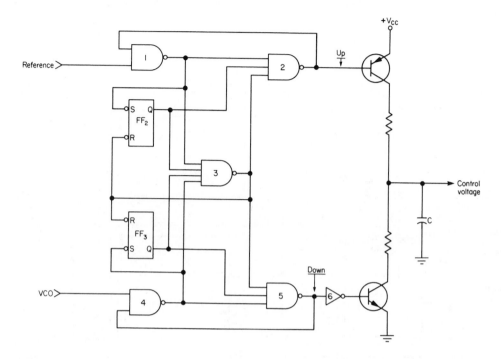

Figure 9.17   Combination phase-frequency detector with "charge-pump" filter at its
output.

showing a difference-frequency sine wave. This will tell the VCO to move to a
higher frequency so that lock can be established. If the reverse is true, the control voltage stays at its lowest level until the VCO comes low enough.

There are several advantages to this. First, the loop cannot lock up on a
harmonic. Second, the capture range of the loop will cover the full operational
range of the VCO. And, finally, the two inputs will always be in phase when the
loop is locked, although this is rarely an important detail. As an added bonus,
the circuit produces an output that indicates whether the loop is in lock or not.

The circuit diagram for this detector is shown in Figure 9.17 and some
sample waveforms are shown in Figure 9.18. Although it looks very complex,
that doesn't really matter. Several manufacturers package the complete circuit
in IC form. Both a TTL (transistor transistor logic) version, that operates off a
5.0 volt power supply, and a CMOS (complementary metal oxide semiconductor) version, that operates from any supply between 5 and 15 volts, are available. (The CMOS version uses NOR gates instead of NAND.)

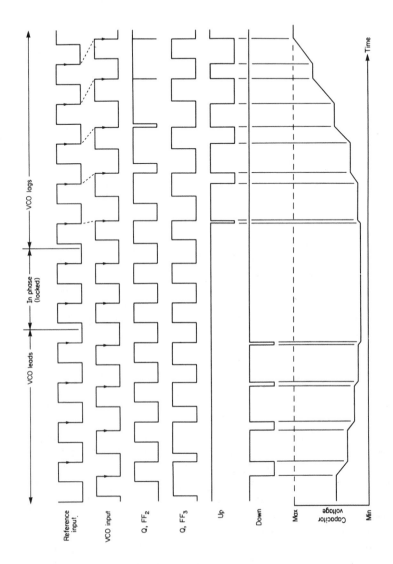

Figure 9.18 Input and output waveforms for the combined frequency-phase detector.

The circuit consists of four SET-RESET flip-flops that are made up of gates 1 and 2, flip-flops #2 and #3, and gates 4 and 5. Gate 3 generates a RESET pulse when all the inputs are Hi. The two output transistors are normally off so that no current will flow through either one to the filter capacitor. This would mean that the two points labeled *UP* and *DOWN* are normally Hi.

The two inputs, VCO and REFERENCE, are shown as symmetrical square waves although, for this comparator, they could be pulses of almost any width. The phase comparison is made on the falling edge of the two inputs. (For the CMOS version, the comparison is made on the two rising edges.) In the sample waveforms, the REFERENCE input is shown with a constant frequency as would be the case in most applications. The VCO input is shown initially with a higher frequency. It slowly decreases until it has the same frequency and phase as the reference for several cycles and then finally it moves to a lower frequency. A point is reached where the VCO signal "appears" to be leading once again but, if a count of actual cycles is made, it can be seen that the lag now exceeds one full cycle of 360°. Although a lag of more than 360° is often referred to as a *leading angle* for simplicity, the difference is important with this detector since some memory is included in the circuit's operation. This means that if, for some reason, the VCO abruptly missed one full cycle but then immediately came back with the right frequency and phase, the detector would know what happened and so momentarily increase the VCO frequency to make up for the missing cycle.

For the short period of time that both input cycles match, the UP and DOWN points remain Hi. Both output transistors stay off so that no charge is added to or taken away from the capacitor. When the VCO leads slightly, the DOWN output pulses Lo for a time proportional to the phase difference. At each pulse, the NPN transistor conducts, causing a momentary path for current to flow to ground. The capacitor voltage will, therefore, drop a small amount each time. When the VCO lags, the UP output pulses the PNP transistor on and off and this will cause the capacitor voltage to rise a small amount each time. The capacitor voltage and the phase detector will always settle at a point where the two input frequencies *and phases* are identical.

The filtering that is happening here is a bit different from the standard RC filters used with the previous phase detectors. Those filters were always connected to a low-impedance source and the source was constantly going Hi and Lo. The filters in that case were having to create an average of the alternating waveform as shown in Figure 9.19. With the more exotic phase-frequency detector, the source is a constant current that can be turned on and off and reversed in direction. The capacitor voltage can be increased or decreased or left alone. The result is that the capacitor voltage is the "integral with respect to time" of the phase differences.

If the VCO frequency is consistently low and its phase keeps lagging more and more so that the angle eventually exceeds 360°, it is important to note that the UP output keeps pulsing in an attempt at raising the capacitor voltage and

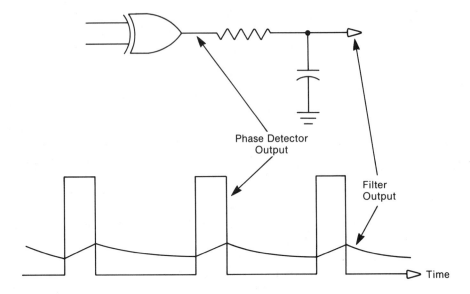

Figure 9.19 Low-pass filters connected to conventional phase detectors will be constantly charging and discharging by small amounts in an attempt to form the average of the phase detector's output.

increasing the VCO's frequency. Simpler comparator circuits would have alternately tried to raise and then lower the VCO frequency as the phase "appeared" to lead and then lag. This problem limits the capture range of the simpler detectors and allows the frequency-phase detector to control the full VCO range without danger of locking on harmonics.

When the loop is out of lock, either the DOWN or the UP point will be pulsing Lo. A two-input NAND gate can be connected to these two points to indicate the out-of-lock condition.

9.12 FREQUENCY SYNTHESIS

Now we examine the most elaborate of PLL applications—the frequency synthesizer. The synthesizer circuitry attempts to generate a frequency with the stability and accuracy of a quartz crystal but with the added convenience of being tunable, in steps, over a wide frequency range.

An example would be a synthesizer that could produce a sine wave at 1.0 MHz or 1.1 or 1.2 or 1.3 etc., right up to 10.0 MHz with increments of 100 kHz all the way. Each frequency would have the same stability and accuracy that individual crystals would have provided but without the bulk and expense of 91

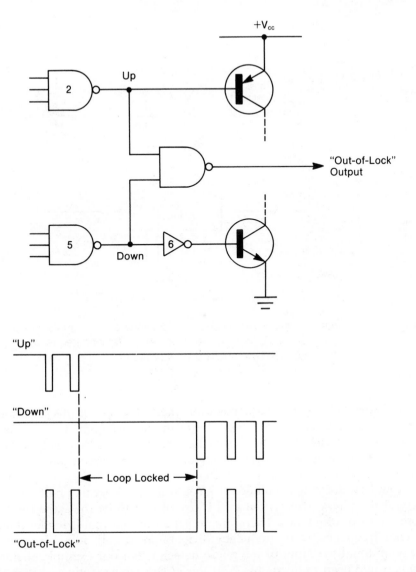

Figure 9.20 Out-of-lock detector consists of a single NAND gate added to the UP and DOWN outputs of the phase/frequency detector. Its output will remain low as long as the loop is locked.

crystals. The PLL version would require a small number of integrated circuits, a selector switch, and a single crystal. A block diagram of such a frequency synthesizer is shown in Figure 9.21.

The synthesizer does not produce a continuous range of frequencies the same as a manually tuned LC oscillator would. Rather the output "steps" in

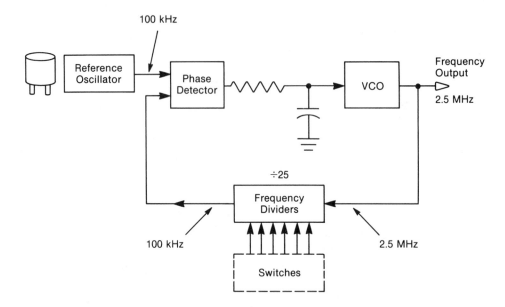

Figure 9.21 A frequency synthesizer consists of the standard PLL components plus some programmable frequency dividers. The reference frequency will always be supplied by a quartz crystal oscillator.

small increments from one frequency to the next. The size of this step is determined by the reference signal. For the example just quoted, the reference would have to be 100 kHz. The circuit then effectively picks a multiple of this input. For example, if a 2.5 MHz output is desired, the circuit must effectively multiply the reference frequency by 25.

Now, how does the circuit do this? You will remember that both the inputs to any of the phase detectors in a PLL circuit had to be the same frequency before the loop would lock. Therefore, to get an output that is higher than the reference frequency, some frequency dividers are simply added between the VCO output and the input of the phase detector. For our example, the added circuits would need to divide by 25. If the VCO wasn't exactly at 2.5000 MHz, the input of the dividers wouldn't be at exactly 100.00 kHz and the loop would act to correct the discrepancy. To switch to a new frequency, the divider circuit would be altered to divide by a new value. There would then be a short period of time when the loop was out of lock until the VCO was pulled to the new frequency.

As mentioned, this circuit is simply creating a harmonic of the reference frequency. But we have already looked at frequency multiplier circuits that accomplished the same thing without all the elaborate circuitry. What is the difference?

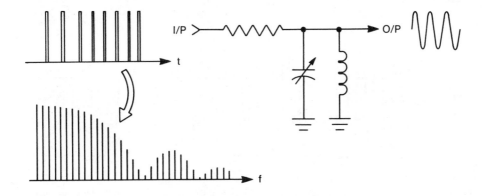

Figure 9.22 A possible alternative to PLL frequency synthesis.

As a possible alternative to the PLL approach, consider the circuit in Figure 9.22. An oscillator can be built that will generate very narrow pulses at a 100 kHz frequency. This pulse shape will have a very strong harmonic content, as shown in the frequency spectrum of the same diagram. All that is needed is a tunable bandpass filter to pick out the harmonic we want and the circuit will be just as good as a PLL circuit—or will it?

The practical problems arise as soon as we look at the loaded Qs required to do the job. If we start with a loaded Q of 10 as an example, we run into the situation shown in Figure 9.23.

If the third harmonic is selected, the filtering is quite good and the resulting output will be a high-quality sine wave. But, at the 16th harmonic, several others above and below the desired one will also pass through and the resulting output will be less than perfect. The resulting output is shown in Figure 9.24.

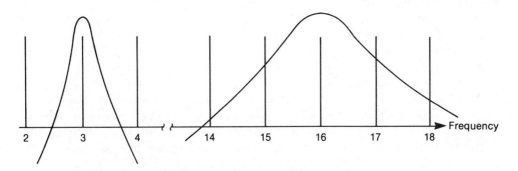

Figure 9.23 A bandpass filter with a constant Q will allow more than the desired harmonic to pass through when high multiples are selected.

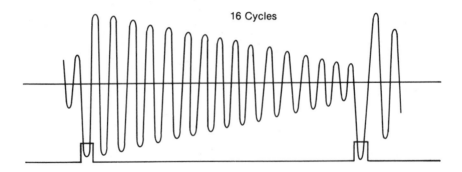

Figure 9.24 The 16th harmonic output. Some higher and lower harmonics have also been included.

A better looking wave could be obtained if the loaded Q was made higher. A value of about 50 would provide the same results as a Q of 10 did for the third harmonic. But there is a limit to just increasing the Q, especially if the 90th harmonic at 9.0 MHz were desired and a loaded Q of about 300 would be needed. But, even if this were practical, another problem arises. Especially with the higher harmonics, it would be very difficult to tune to the exact value desired. The 89th harmonic, for example, would be a very small "twist of the dial" away from the 90th harmonic.

Therefore, harmonic filtering is practical only for small multiples with a practical upper limit of about 5.0. True frequency synthesis belongs to the PLL approach.

This short look at a limited alternative to the PLL can show us something else. Figure 9.24 effectively showed that the tuned circuit was provided with energy every 10 microseconds (the period of the 100 kHz pulses). Between pulses, it is up to the tuned circuit to keep the oscillations going. The figure shows them slowly dying out as the energy is used up. (The amount of stored energy is directly related to the Q of the circuit.) If the self-resonant frequency is slightly off, the frequency of oscillations between input pulses will not be an exact harmonic. There would also be an abrupt phase change each time the next input pulse appeared.

The PLL can suffer from somewhat similar problems but they aren't nearly as severe. The phase and frequency of the VCO are really only being compared once every cycle of the reference frequency. Between comparisons, the VCO can do whatever it wants. However, remember that, because of the low-pass filter, the "average" frequency and phase are being controlled, not the instantaneous frequency, once every 10 microseconds. Second, the oscillator will always hold its amplitude constant (there may be small random variations due to noise) so the output will not decay between reference pulses. The end result is that the PLL synthesizer will work extremely well as long as the oscillator is capable of behaving itself over a short period of time. The loop feature will look after the average frequency and phase and eliminate any slow drifts, but the more rapid frequency, phase, and amplitude jitter must be eliminated by designing the VCO properly in the first place.

9.13 FREQUENCY DIVIDERS

The digital dividers must take the higher frequency from the VCO and divide it down to match the lower reference frequency. The division ratio is set by the position of some type of input switches that are used by the operator to "dial up" the desired frequency. Some type of numeric values, either on the switch itself or on a separate light emitting diode display, will indicate the actual frequency that results once the loop locks.

The counters are called *programmable* since their dividing ratio is determined by the external switches. Programmable counters are made to recognize two different binary inputs—BCD (binary coded decimal) and normal binary. BCD is a four-bit code that represents only the numerals 0 through 9.

Normal binary can represent any number as long as enough control lines (bits) are used; four lines could represent 0 to 15, five lines 0 to 31, and six lines 0 to 63 (typical of 40-channel CB synthesizers). The higher numbers in BCD are accommodated by adding four more control lines for each digit needed.

The internal logic circuitry consists of a set of down counters that can be initially preset to any number. To understand the operation of the program-

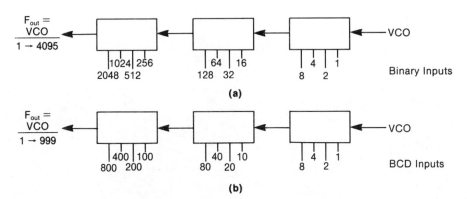

Figure 9.25 Programmable frequency dividers for (a) binary input, and (b) BCD input. Since division by 0 has no meaning, the smallest division is 1, i.e., the input and output frequencies are the same.

mable counters and how they operate as part of the loop, we will consider the following example.

A synthesizer is required to produce 99 frequencies spaced 50 KHz apart with the first frequency 50 KHz. The highest frequency would be 99 × 50 = 4.95 MHz. (Designing the VCO for this 100:1 range would be interesting but we are looking at the dividers at the moment.) The programming switches and, therefore, the counter types will be BCD. The circuit could look like Figure 9.26. For this example, we will consider that the VCO is required to operate at 25 times the reference frequency.

The thumbwheel switches provide the BCD code for 2 (0010) and 5 (0101). When the cycle starts, the flip-flops in the counter are loaded (preset) with the 0s and 1s from the switches. The square wave pulses from the VCO (at 1.25 MHz) start the least significant counter downward. The first VCO pulse clocks the counter down to:

4 (0100)
then 3 (0011), VCO pulse #2
" 2 (0010), " pulse #3
" 1 (0001), " pulse #4
" 0 (0000), " pulse #5

The sixth VCO pulse sets this counter back to 9 (the switch inputs are ignored). Up until this time, the second counter, which was preset to 2, doesn't change. As soon as the first counter goes back to 9, the second counter is provided with the appropriate clock pulse and counts down by 1. The first counter now sits at 9 and the second at 1. Six pulses from the VCO have been used up. The count continues:

Second counter (0001), First counter 9 (1001), VCO pulse #6
" 8 (1000), " " #7

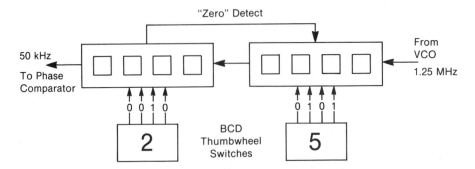

Figure 9.26 Two BCD counters are programmed by thumbwheel switches to divide the VCO frequency by 25.

"	7 (0111),	"	"	#8
"	6 (0110),	"	"	#9
"	5 (0101),	"	"	#10
"	4 (0101),	"	"	#11
"	3 (0011),	"	"	#12
"	2 (0010),	"	"	#13
"	1 (0001),	"	"	#14
"	0 (0000),	"	"	#15

The 16th VCO pulse again sets the first counter back to 9 and again the second counter counts down by 1. The total situation now sits at:

Second counter (0000), First counter 9 (1001), VCO pulse #16

The count continues:

(0000)	8 (1000)	"	#17
	7 (0111)	"	#18
	6 (0110)	"	#19
	5 (0101)	"	#20
	4 (0100)	"	#21
	3 (0011)	"	#22
	2 (0010)	"	#23
	1 (0001)	"	#24
	0 (0000)	"	#25

On the 25th VCO pulse, both counters now contain all 0s. This situation is sensed by a "zero detector circuit" and presets the counters immediately to 2 and 5. Only now does an output pulse get passed on to the phase comparator. We have come back to our starting point; 25 VCO pulses have been used and only one was fed to the phase comparator. The total circuit has, therefore, divided the VCO frequency by 25, the same number dialed on the thumbwheel switches.

9.14 REVIEW QUESTIONS

1. The total circuitry involved in a phase-locked loop is fairly complex. Why is the technique so popular?

2. What is the purpose of the low-pass filter in the loop?

3. What is the "gain" of a VCO that shifts from 4.1 to 11.6 MHz when the DC control voltage is adjusted from 2.1 to 5.6 volts? Be sure to include the correct units.

4. What is the "gain" of the exclusive-OR phase detector (Figure 9.13)?

5. Find the DC voltage that will appear at the output of a simple low-pass filter connected to the output of an exclusive-OR phase detector. The two inputs are at the same frequency but are 115° out of phase. Assume that a Logic Hi from the gate is 3.6 volts and a Lo is 0 volts.

6. What is the 3 dB cutoff frequency for a low-pass filter consisting of a 47 k ohm series resistor and a 0.47 μF parallel capacitor?

7. The two inputs to an exclusive-OR phase detector are 23 kHz and 21.8 kHz. The simple low-pass filter connected to the output has a 3 dB cutoff frequency of 75 Hz. Assuming the same logic levels as in question 5, what is the average DC voltage at the filter's output and what is the peak-to-peak amplitude of any AC component?

8. What is the main difference between the measurement for tracking range and capture range?

9. A PLL by itself could be used to make a very simple FM receiver but not an AM receiver. Why?

10. Compare the phase of the two input signals in a locked loop if an exclusive-OR phase detector is replaced with a phase-frequency detector.

11. For a frequency synthesis of 4.0, 4.2, 4.4 . . . 5.0 MHz, what must the reference frequency be? What values must the frequency dividers be set to?

12. What frequency will be synthesized if a reference frequency of 100 kHz is used and the feedback dividers are set to divide by 435?

13. Does the quality of a synthesized frequency deteriorate in any way when it is a high multiple of the reference frequency?

14. Compare the advantages and disadvantages of binary and BCD frequency dividers.

15. Sketch a block diagram for a synthesizer circuit to produce frequencies of 5.00, 5.01, 5.02, 5.03 . . . 5.99 MHz. The dividers will be BCD-type.

10. TRANSMISSION LINES

This chapter is concerned with those "long" connecting wires that carry "high-frequency" signals from one place to another. These wires might form the long telephone lines used for voice or data communications, or that long antenna cable for your television receiver. But they might also be the set of short copper traces on a PC board that connect microprocessor chips with a "bus." Some connecting paths may not even be wires. Hollow metal tubes called *waveguides* and thin optical glass fibers also form very efficient transmission lines.

Figure 10.1 shows some variations of long transmission lines and also shows a much shorter application illustrating that lines of almost any length may, on occasion, have to be analyzed as transmission lines.

10.1 WHAT IS A TRANSMISSION LINE?

The correct answer to this question is—every electrical connection, wire, cable, etc., that connects a signal source to a load. The problem with this answer is that it means that all calculations involving even the simplest piece of wire will be fairly complex because transmission line theory isn't a simple extension of Ohm's law. Therefore, we look for a simpler alternative whenever possible. In many cases, when a wire isn't too long, we need consider only the equivalent inductance and capacitance of the wire for simplicity. For very short lengths, such as for the leads on a resistor, we may even (fortunately) be able to totally ignore any effects.

The ultimate answer is that a pair of wires must be considered as a transmission line when all other approaches fail—when simple calculations won't match what is observed in practice. By the end of this chapter, we will have developed a more definite answer based on "impedance matching," line length, and operating frequency.

Now, when we actually have to consider a set of connecting wires as a transmission line, what is happening?

- Current traveling along any wire creates a magnetic field around that wire. This field stores energy and may overlap other wires or objects that are close. This property gives wire its *inductance*.

- Voltage between two wires will create an electric field around the wires. This field also stores energy and may also overlap other wires and nearby objects. This property creates a *capacitance* between wire pairs.

- Any signal traveling along a wire pair takes a very definite amount of time to travel the full length. The longer the wires, the more time is taken.

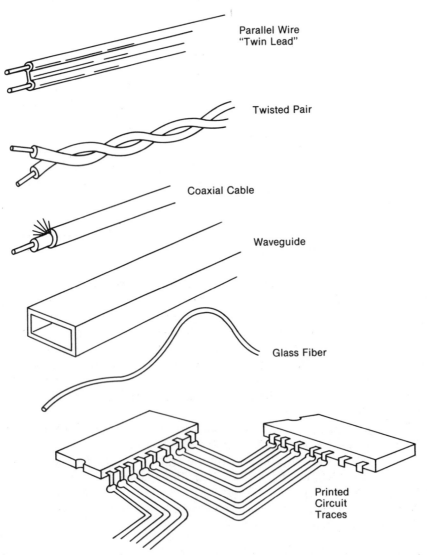

Figure 10.1 Transmission lines are normally thought of as long wires, tubes, or fibers. But, in some cases, even short paths carrying high-frequency signals must also be considered as transmission lines.

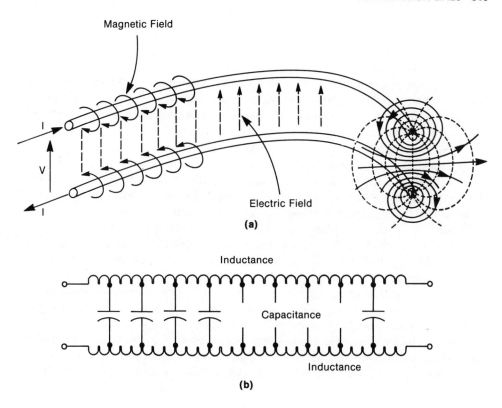

Figure 10.2 The two fields that surround a pair of wires as a result of the current through the wires and the voltage between them (a). The equivalent electrical circuit contains distributed inductance and capacitance (b).

This last point ends up complicating the first two and is the real reason that transmission lines aren't handled like any other circuit. The traveling time prevents us from "lumping" all the inductance of the entire cable into one equivalent component and all the capacitance into another equivalent component. (For short lines this may be possible and so saves us the bother of the more complex situation.) Instead, the traveling time forces the inductances and capacitances to be spread out or "distributed" along the length of the line. Figure 10.2 shows the two fields that surround a pair of wires and the distributed circuit equivalent of the line.

10.2 ELECTROMAGNETIC FIELD

If you look at Figure 10.2 once again you will notice that the electric and magnetic fields always cross at right angles. This creates a very special situation called an *electromagnetic field*. When we study antennas and propagation, we will see that this is a wave that can also exist by itself and travel through free space without wires.

The voltages, the current, and the electromagnetic field interact very closely. When the field follows the wires and if something alters the shape or strength of the field, the voltages and currents will be affected. Or, if something changes a voltage or current, the field will be affected.

The field surrounds the wires but the strongest portion is between the two conductors. In many commercial cables, this is the area of insulation. Therefore, we end up with the strange situation that the characteristics of the cable depend on the type and amount of insulation between the wires. Actually, we shouldn't be too surprised because the value of a capacitor depends on the type of insulation between its plates and we have the same situation here.

The property of the insulation that is most important to us is it's *dielectric constant (k)*. This is a value that compares the dielectric properties of the material with that of free (empty) space. This number describes how hard it is to set up an electric field inside the material. A value of 2, for example, means that for a given voltage between two wires, the electric field will only be half as strong in that material as it would have been in free space.

The velocity of the electromagnetic signal along (and between) the wires depends on the dielectric constant of the insulation. The larger this number, the slower the signal travels. The relation between velocity and dielectric constant is:

$$\text{velocity (meters/sec)} = \frac{300 \times 10^6}{\sqrt{k}} \qquad (10.1)$$

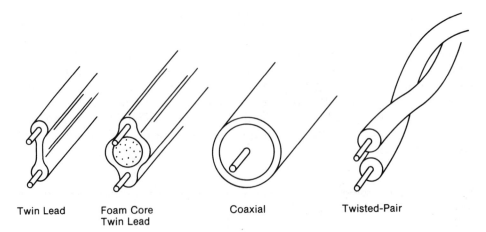

Twin Lead   Foam Core   Coaxial   Twisted-Pair
                   Twin Lead

Figure 10.3 End views of four typical transmission lines showing different types of insulation (dielectrics) between and around the conductors.

where k is the dielectric constant of the insulation. It is a
relative value (no units).

Some typical dielectric constants and resulting line velocities are given in
Table 10.1. Notice that the value for free space is 1.000 and the resulting speed
is that of light.

Table 10.1 Dielectric constants and velocity of electromagnetic waves
for typical insulating materials.

Material	Dielectric Const. (k)	Velocity (meters/sec)
Vacuum	1.000	300×10^6
Air	1.0006	299.9×10^6
Teflon	2.10	207×10^6
Polyethylene	2.27	199×10^6
PVC	3.30	165×10^6
Nylon	4.90	136×10^6
Polystyrene	2.50	190×10^6

While velocities and insulating materials are being discussed, it is worthwhile to consider some potential problems. On most two-wire lines, the field
can extend well beyond the limits of the dielectric. Any clips, nails, or tape that
are being used to fasten down the wires will, therefore, affect the passage of the
wave to some extent and so must be considered as potential problems. The fact
that some of the wave travels in the insulation and some travels in the air also
means that two different velocities are involved. For some very special applications involving long lines at high frequencies, this could present a problem. A
weak signal could reach the end of the cable first using the higher-speed path—
in the air. Then, a while later, the main signal would arrive—traveling through
whatever type of plastic or rubber made up the insulation. Coaxial cable avoids
both of these problems because the signal is completely enclosed by the outer
conductor.

10.3 WAVELENGTH

As a high-frequency signal travels along a transmission line, it is also oscillating·
through its sine wave cycle. If a "snapshot" of all the voltages along the line at
one instant could be taken, you would observe one full electrical cycle "spread
out" over some distance on the line. This distance is called the *wavelength*. The
physical length will always be equal to or shorter than the wavelength of the
same signal in free space. The amount will depend again on the dielectric con-

stant and the resulting velocity. The slower the velocity, the shorter the wavelength of any given frequency. The relationship is:

$$\text{Wavelength (meters)} = \frac{300 \times 10^6}{\sqrt{k} \times \text{frequency (Hz)}} \qquad (10.2)$$

Example 10.1

What is the wavelength of a 20.7 MHz sine wave on a transmission line that uses polyethylene insulation?

Solution:

From Table 10.1 we find that the dielectric constant *(k)* of polyethylene is 2.27. Using this we find that the wavelength on the line is:

$$\text{Wavelength} = \frac{300 \times 10^6}{\sqrt{k} \times \text{frequency}}$$

$$= \frac{300 \times 10^6}{\sqrt{2.27} \times 20.7 \times 10^6}$$

$$= 9.62 \text{ meters}$$

10.4 EQUIVALENT CIRCUIT

To proceed any further with a description of transmission lines, it is necessary to start considering the inductances and capacitances of the wires. But this looks impossible since, as we saw back in Figure 10.2, the components are distributed all along the length of the line. There are two steps that take us out of this dilemma. The first should seem reasonable, the second probably won't—at first.

1. We can measure the equivalent components of a line as long as we use a sample length short enough to remove any significant traveling time effects.

2. Any calculation performed on the sample length can be extended to any length at any frequency.

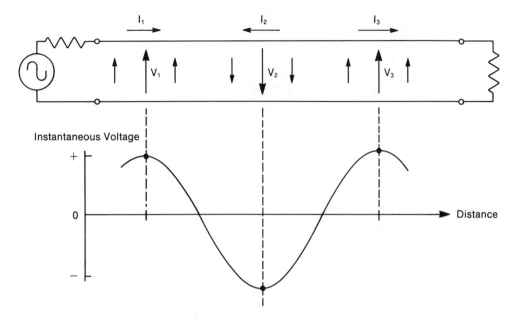

Figure 10.4 At any instant in time, the currents along the wires and the voltages between the wires are not all the same. The physical distance occupied by one full electrical cycle is called *one wavelength* (λ). Wavelength depends on frequency and insulation type.

Now back to the first step. If we have a capacitance/inductance meter of some type it is easy to look in its manual and find out what frequency it uses to make its measurements. For our example, we will assume it is 10 kHz. The sample length of cable should be cut at 1/360th of a wavelength or less at this frequency. We will assume a dielectric constant for our example of 1.00.

$$\text{Wavelength} = \frac{300 \times 10}{\sqrt{1} \times 10 \text{ kHz}} = 30{,}000 \text{ meters}$$

$$1/360 \text{ th wavelength} = 83.3 \text{ meters}$$

By keeping the sample length less than this, there will be very little problem with traveling time since the current entering the cable will be less than one degree out of phase with the current leaving.

Figure 10.5 shows the two measurements. First, the capacitance can be measured between the two wires at the one end of the cable with the other end open. Then the end-to-end inductance of one of the wires can be measured. The resulting equivalent circuit is shown in the same figure. Notice that the mea-

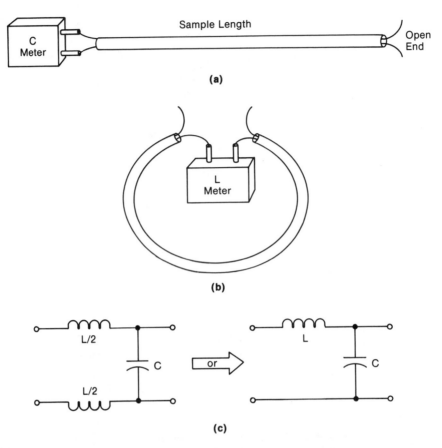

Figure 10.5 Equivalent circuit (c) for a sample length of transmission line. The measurement setups are shown in (a) for the capacitance and (b) for the inductance of one of the wires.

sured inductance value is used twice in the equivalent circuit and counts for one-half of the total inductance. (One wire was measured and there are two wires in the cable.)

It must be remembered that these component values were obtained for a rather arbitrary length of line and so we must be careful what assumptions are made for the longer line. For example, the equivalent circuit looks very much like a low-pass filter and one might be tempted to calculate resonant frequencies and cutoff frequencies. While it is true that a line can be resonant, we will see shortly that this is more a function of its physical length than its equivalent circuit values. The idea of a cutoff frequency also cannot be used because a shorter sample length would have smaller equivalent values and so might seem to have a higher cutoff frequency. The losses of any practical line do increase

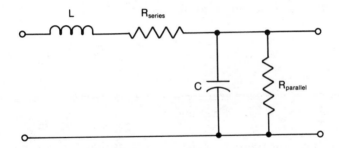

Figure 10.6 A more thorough measurement of the equivalent circuit for a sample length of transmission line shows resistance components that create losses. The series resistance shown is due to the skin effect in the wires and so will increase with frequency. The parallel resistance comes from insulation losses that are more or less constant at most frequencies. All the inductance has been combined into a single value.

with frequency but this is a result of increases in the "skin effect" of the wires and insulation losses and has nothing to do with inductance and capacitance.

One possible wrong conclusion might be that two one-meter lengths of line soldered together would have a different set of characteristics than a single two-meter length and, of course, this isn't the case. Full-length transmission line calculations are obtained by considering the final answer that calculations seem to be converging on when a large number of sample lengths are connected end-to-end. This type of calculation is an excellent job for calculus. We won't go into the formula development here; just the end results will be presented.

10.5 VELOCITY

Using the equivalent components, the formula for the velocity of the wave along a transmission line is given here. Its form should seem reasonable because an increase in dielectric constant would raise the capacitance and so lower the velocity. This is just what our previous statement based on dielectric constant alone told us. The formula given is for a line with no losses. This will suit most applications, partly because a full formula that includes losses is much harder to handle and partly because the velocity changes that do occur because of the losses are minor. For lossless lines:

$$\text{Velocity (meters/sec)} = \frac{1}{\sqrt{LC}} \qquad (10.3)$$

where l = sample length in meters
C = capacitance of the sample length in farads
L = total inductance of both wires, in henries.

10.6 CHARACTERISTIC IMPEDANCE

Because of the finite traveling time, the signal traveling forward on the line has no way of knowing what is at the other end. And so, the voltages between the wires and the currents through the wires set themselves up in a relationship that is called the *characteristic impedance* (Z_0) of the line.

$$\text{Characteristic impedance } (Z_0) = \frac{\text{Forward Voltage}}{\text{Forward Current}} \text{ (ohms)} \qquad (10.4)$$

This value is mainly resistive but doesn't represent any real loss of energy. It is simply the ratio of the voltages and currents traveling in any one direction.

The condition is similar to driving your car along a narrow twisting road at night. You would tend, hopefully, to drive according to the local conditions of the road instead of thinking of what is ahead and speeding up or slowing down.

Can characteristic impedance be measured? Yes. One ridiculous method is to use an ordinary ohmmeter. When the leads are clipped across the one end of a long pair of wires (open at the other end), the meter will initially display the characteristic impedance. But the reading will only last for a short time until the energy from the meter's battery has traveled all the way to the end and then come back because there was no place else to go. Then the meter would show an open circuit.

While this method would work, the time interval for anything except super-long wires would be too short to get a reading. The same idea does work if a narrow pulse is used and the "resistance" is measured during this short time interval using a high-speed voltage and current measurement.

Characteristic impedance can also be calculated if the equivalent components of a sample length are known. Because the calculation involves the ratio of the two reactances, the impedance does not depend on the sample length. A piece twice as long would have double the capacitance and inductance so the calculation would give the same answer for the impedance. For a lossless line:

$$Z_0 = \sqrt{L/C} \text{ (ohms)} \qquad (10.5)$$

For a line with significant losses:

$$Z_0 = \sqrt{\frac{R_s + j2\pi L}{(1/R_p) + j2\pi fC}} \text{ (ohms)} \qquad (10.6)$$

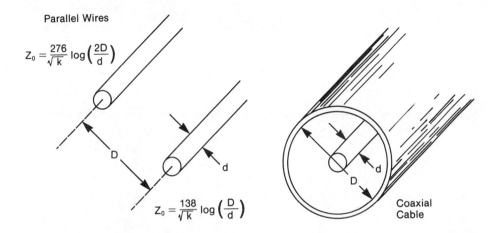

Parallel Wires

$$Z_0 = \frac{276}{\sqrt{k}} \log\left(\frac{2D}{d}\right)$$

$$Z_0 = \frac{138}{\sqrt{k}} \log\left(\frac{D}{d}\right)$$

Coaxial Cable

Figure 10.7 Characteristic impedance of open-wire and coaxial lines can be calculated when the physical dimensions and dielectric constant (k) are known.

This is an important formula because it says that the characteristic impedance will change with losses. The series loss due to skin effect will be the most serious so there may be some impedance changes with frequency in a lossy cable.

The characteristic impedance can also be determined strictly from the physical properties of the line. Figure 10.7 shows the calculation for an open-air line and also for a coaxial line that uses a dielectric material.

Example 10.2

What is the characteristic impedance of a parallel wire transmission line if the inductance of "each" wire is 0.25 μH per meter and the capacitance between the lines is 30 pF per meter?

Solution

The inductance was given for only the one wire so the value must first be doubled to give the total inductance for the sample length.

$$L = 0.50 \ \mu H$$
$$C = 30 \ pF$$

$$\text{Characteristic impedance} = Z_o = \sqrt{L/C}$$

$$= \sqrt{\frac{0.50 \times 10^{-6}}{30 \times 10^{-12}}}$$

$$= 129 \text{ ohms}$$

Example 10.3 ══

What is the characteristic impedance of a coaxial transmission line insulated with polystyrene that has the following dimensions? The diameter of the outer or "shield" conductor is 5 mm and the inner conductor has a diameter of 0.8 mm.

Solution:

From Table 10.1, the dielectric constant of polystyrene is 2.50. The characteristic impedance is then:

$$Z_0 = \frac{138}{\sqrt{k}} \log \left(\frac{D}{d} \right)$$

$$= \frac{138}{\sqrt{2.50}} \log \left(\frac{5}{0.8} \right)$$

$$= 87.28 \times 0.796$$

$$= 69.5 \text{ ohms}$$

10.7 TERMINATIONS AND REFLECTIONS

While a signal is traveling in any particular direction along a line, its voltages and currents assume the ratio known as the *characteristic impedance*. However, when the signals reach the end of the line where the load is connected, they could be in for a shock. If the load resistance happens to be exactly the same as the characteristic impedance, the signal happily moves into the load and is completely used up. But, if the load isn't exactly the same as the characteristic impedance, some of the signal turns back toward the source. The amount that turns back depends on how badly the characteristic impedance and load are "mismatched." The ratio of the forward voltage heading toward the load and the reflected voltage is known as the *reflection coefficient*. Its value can vary from 0 (no reflection) up to 1.0 (total reflection).

$$\Gamma = \frac{V_{\text{reflected}}}{V_{\text{forward}}} \tag{10.7}$$

The relation between the reflection coefficient, the characteristic impedance, and the load impedance is:

$$\Gamma = \frac{Z_{\text{load}} - Z_0}{Z_{\text{load}} + Z_0} \text{ (no units)} \tag{10.8}$$

This could be a complex number if the characteristic impedance and the load impedance are not both purely resistive.

The reflected voltage may undergo an immediate phase shift just as it changes direction. If both the characteristic impedance and the load are resistive and the load is higher than the line impedance, no shift occurs. If the load resistance is lower, a 180° shift occurs. For complex impedances, the shift would be some value between 0 and 360°.

Now we have two waves traveling on the transmission line—a forward one traveling toward the load and, possibly, a weaker, reverse one traveling back toward the source. Let's stop for a moment to make sure we can identify these two.

The two separate waves can only be distinguished if a special piece of test equipment known as a *directional coupler* is used. It has two sets of terminals — one for the forward voltage and one for the reverse voltage.

Each wave travels along the line, paying attention only to the characteristic impedance at that exact spot on the line. The fact that another wave may be passing in the reverse direction doesn't bother the forward wave in the slightest. It's as if the two waves were like cars on a two-lane road and their drivers

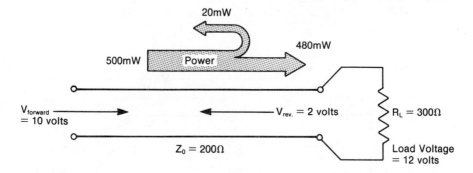

Figure 10.8 When the load resistance isn't the same as the line's characteristic impedance, some signal must turn back.

kept their eyes on the road immediately ahead. However, if a voltmeter or oscilloscope is connected across the line at some point, it has no ability to separate the two waves and so sees only a total voltage. Now there will appear to be additions and subtractions of voltages with results that depend on distance along the line.

As we proceed to the next section, keep these points in mind. The two waves are separate but most measuring instruments cannot distinguish directions. The results are voltage variations and apparent impedance changes that are position-dependent.

10.8 STANDING WAVES

Figure 10.9 shows the interference pattern along the length of the line that can be observed if nondirectional measuring equipment is used. The results are known as a *standing wave pattern.* A common number that is used to describe how bad the reflection is is *voltage standing wave ratio* (VSWR). This is simply

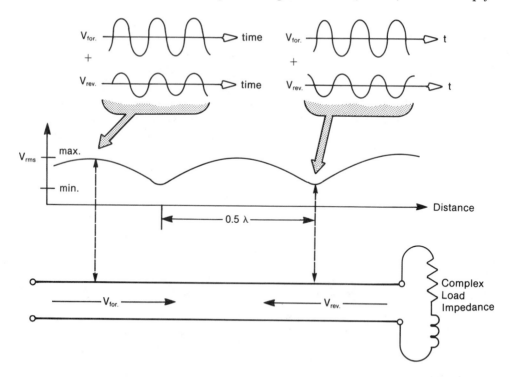

Figure 10.9 A mismatch results in a signal reflection. The magnitude and phase of the reflected signal depend on the type of mismatch and the result is a series of voltage maximums and minimums along the line.

the ratio of the largest voltage on the line to the smallest. With no reflected signal (the ideal case), the VSWR will be 1.0 because there will be no variation in total voltage along the length of the line. The worst case occurs when all of the forward voltage is reflected, causing a total voltage that would change from double the normal value down to 0 as the two waves alternately add and subtract. The resulting VSWR would be infinite. This will happen if the end of the line is open, shorted, or connected to a purely reactive load.

$$\text{VSWR} = \frac{\text{Voltage maximum}}{\text{Voltage minimum}} \text{ (no units)} \qquad (10.9)$$

$$\text{VSWR} = \frac{1 + |\Gamma|}{1 - |\Gamma|} \qquad (10.10)$$

(The vertical bars mean "absolute value" or "magnitude.")

$$\text{VSWR} = \frac{Z_0}{Z_{\text{load}}} \text{ or } \frac{Z_{\text{load}}}{Z_0} \text{ (whichever is greater)} \qquad (10.11)$$

Finally, the messy problem. What effect does all of this have on the input impedance of a transmission line?

The input impedance (for a continuous sine wave) will vary all along the line if there are any reflections from a mismatched load. The average impedance along the length of line will still be the same as the characteristic impedance but the exact value will be higher at some points and lower at others and the phase angle of this impedance will also be changing with position. How much variation there will be above and below the characteristic impedance value will depend on the amount of reflection. The highest and lowest impedances on the line are given by the following:

$$Z_{\text{highest}} = Z_0 \times \text{VSWR} \qquad (10.12)$$

$$Z_{\text{lowest}} = Z_0/\text{VSWR} \qquad (10.13)$$

The general formula needed to calculate the input impedance at any point is given below.

$$Z_{\text{in}} = Z_0 \frac{Z_L + j Z_0 \tan \Theta}{Z_0 + j Z_L \tan \Theta} \text{ (ohms)} \qquad (10.14)$$

where Z_0 = characteristic impedance
Z_L = load impedance
Θ = "distance" from load to input point in degrees. One wavelength = $360°$

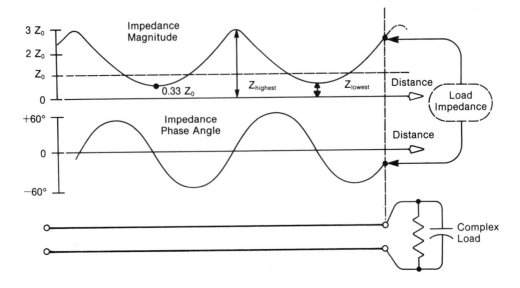

Figure 10.10 The input impedance of a mismatched transmission line varies along its length. The pattern repeats every half cycle. The average impedance is still the characteristic impedance and the amount of variation depends on the line-to-load mismatch. The VSWR is 3.0 in this example.

Example 10.4

Using a situation similar to that shown in Figure 10.8, except that the forward voltage is 5 volts, the characteristic impedance is 125Ω and the load resistance is 100Ω. Find:

- The forward power
- The reverse voltage
- The load voltage
- The load power
- The reverse power
- The VSWR

Solution:

$$\text{Forward power} = \frac{V_{for}^2}{Z_0} \text{ (normal power formula)}$$

$$= \frac{5^2}{125} = 200\text{mW}$$

$$\text{Reflection coefficient} = \Gamma = \frac{Z_{\text{load}} - Z_0}{Z_{\text{load}} + Z_0}$$

$$= \frac{100 - 125}{100 + 125} = -0.1111$$

(The negative sign indicates that a 180° phase reversal will occur at the reflection point.)

$$\text{Reverse voltage} = \Gamma \times \text{Forward voltage}$$
$$= -0.1111 \times 5.0$$
$$= -0.555 \text{ volts}$$

(starts out-of-phase with forward voltage)

$$\text{Load Voltage} = \text{forward voltage} + \text{reverse voltage}$$

$$= 5.0 - 0.555$$

$$= 4.445 \text{ volts}$$

$$\text{Load Power} = \frac{(V_{\text{load}})^2}{Z_L} \text{(normal power formula)}$$

$$= \frac{(4.445)^2}{100}$$

$$= 197.6 \text{ mW}$$

$$\text{Reverse Power} = \frac{(V_{\text{reverse}})^2}{Z_0}$$

$$= \frac{(0.555)^2}{125}$$

$$= 2.4\text{mW}$$

$$\text{VSWR} = Z_0/Z_{\text{load}}$$

$$= 125/100$$

$$= 1.25$$

Example 10.5

What is the input impedance of a line 0.35 wavelengths long? Its characteristic impedance is 100Ω and the load impedance is $75 + j15\Omega$.

Solution:

The "length" of the line in degrees is:

$$\Theta = 360 \times 0.35$$
$$= 126 \text{ degrees}$$
$$\tan \Theta = -1.376$$

$$Z_{in} = Z_0 \frac{Z_L + j Z_0 \tan \Theta}{Z_0 + j Z_L \tan \Theta}$$

$$= 100 \times \frac{(75 + j15) + j100 \tan 126°}{100 + j (75 + j 15) \tan 126°}$$

$$= 100 \times \frac{75 - j122.64}{120.65 - j103.2}$$

$$= \frac{100 \times 143.755 \ \underline{/-58.55°}}{158.766 \ \underline{/-40.54°}}$$

$$= 90.54 \ \underline{/-18.01°} \text{ ohms (polar form)}$$

$$= 86.1 - j27.99 \text{ ohms (rectangular form)}$$

10.9 RESONANT LINES

A special case occurs when the end of a transmission line is either shorted or left completely open. Then the line will show a variation of pure reactances along its length and will include resonant points that are the equivalent of series and parallel resonance in conventional circuits. Resonant lines can be used for tuned circuits in VHF, UHF, and microwave transmitters and receivers. The only reason they aren't normally used at lower frequencies is the increased length of the line.

Figure 10.11 describes the variation of this reactance for the two cases. With the end left open, the reactance appears capacitive for the first-quarter wavelength as you move back from the open end and finally reaches a very low impedance point that is the same as series resonance. Then, as the distance increases further, the reactance increases and is inductive. At a point one-half wavelength from the open end, the impedance is infinite. The same pattern continues to repeat every half wavelength.

If the line is shorted at the end, the reactance is initially inductive as you move away from the end. At a point one-quarter wavelength away, the impedance is infinite. From there the reactance falls again and is capacitive. The shorted line is similar to the open line from this point on.

The "resonance" of shorted and open transmission lines suffers from the same Q problems that ordinary tuned circuits do. The problem is the small losses within the line. However, if large diameter wire and silver plating are used, $Q's$ well over 1000 can be achieved.

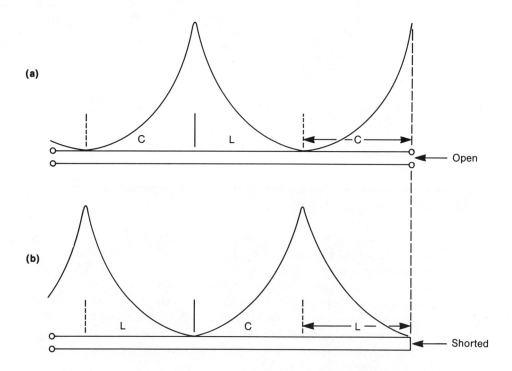

Figure 10.11 Impedance variation along a line that is (a) open at its end and (b) shorted at its end.

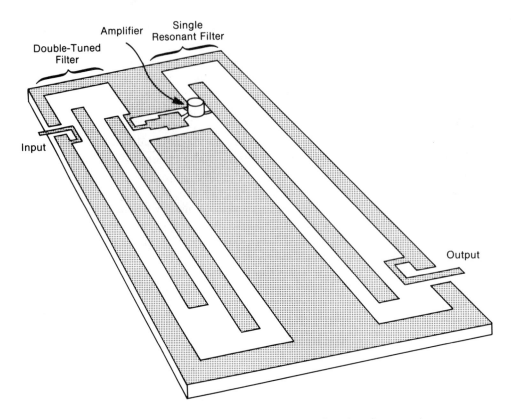

Figure 10.12   Printed circuit bandpass filters using short lengths of resonant transmission lines on a printed circuit board.

10.10   QUARTER WAVE MATCHING

For maximum power transfer and freedom from impedance variations, it is normally desirable to match the load to the characteristic impedance of the line. When operating over a wide frequency range, a transformer is normally needed. An excellent example of this is the 300Ω to 75Ω transformer used to match television coaxial cable to the antenna and the TV receiver. The frequency range, in this case, is 50 to 900 MHz. With television, matching not only improves the power transfer but it also reduces the amount of signal reflections within the cable system. These reflected signals may bounce around off several mismatches and reappear at a later time as a ghost.

For some applications where the bandwidth is much narrower (about 10% of the center frequency), a simpler matching device can be used. This consists

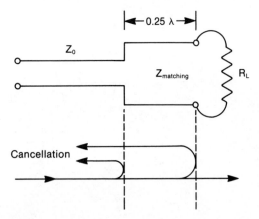

Figure 10.13   Quarter wave matching is useful when small bandwidths are involved. The reflection off the second impedance mismatch cancels the reflection off the first.

of an extra quarter wavelength of line connected between the load and the end of the cable. The characteristic impedance of this extra line is chosen to be the geometric mean of the main line and the load. The idea is that there will now be two reflected signals of the same strength but, because they are one quarter wave apart physically, the second signal will have had to travel a half wavelength further and so be out of phase and will cancel the first signal.

$$Z_{matching\ section} = \sqrt{Z_0 \times Z_L'} \qquad (10.15)$$

10.11   SMITH CHART

For accurate calculations of input impedance on mismatched transmission lines, the formula presented previously (Equation 10.4) must be used. However, it is a tedious calculation and, even with a programmable calculator or computer, does not provide a good working tool. The chart about to be described provides a much more visual (but less accurate) picture of the impedance and any added matching networks.

Developed by P.H. Smith in 1944, the chart is simply a plotting space for all impedances from zero to infinity and with any reactance—capacitive, inductive, or purely resistive. The big advantage of this form of an impedance chart is that the variation of impedances along a mismatched transmission line form simple circles with centers at the mid-point of the chart.

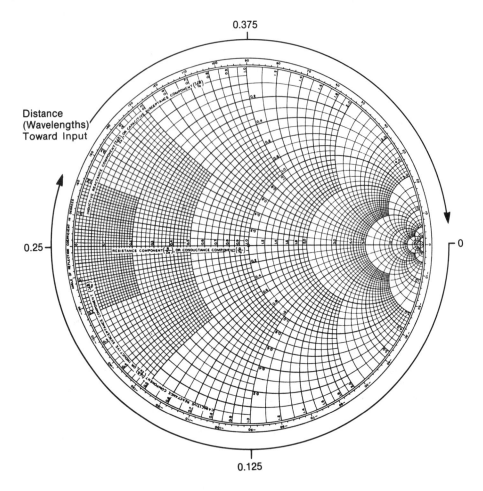

0.375

0.25

0.125

0

Distance
(Wavelengths)
Toward Input

Figure 10.14 The normalized Smith Chart. The center point represents the characteristic impedance of the transmission line being used. Moving away from this center point describes other complex impedances that can exist on a mismatched line.

To make the chart more general purpose, it is normally printed in a "normalized" version although special versions can be purchased for use with 50Ω or 75Ω cable. In any case, the center point of the chart represents the characteristic impedance of the cable being used. In the normalized version shown, this is marked as 1.0 on the center axis. A load resistor that is twice the cable impedance (i.e., 150Ω load on 75Ω cable) would be represented as a dot at the 2.0 point on the center (pure resistance) axis. All the possible impedances (in normalized form) that would appear at different points along the cable can be

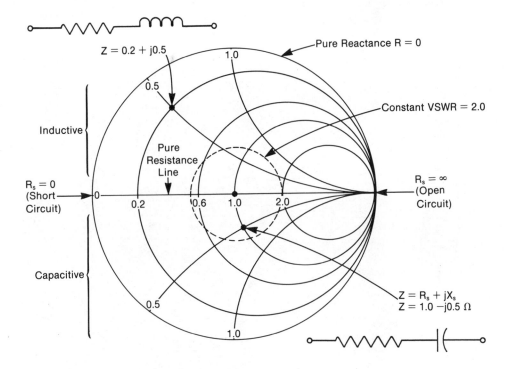

Figure 10.15 Impedance points on the Smith Chart.

found by drawing a circle with its center at 1.0 in the middle of the chart and its circumference passing through the dot that represents the mismatched load. Distances from the load to the desired point on the line are found by moving clockwise around the chart using the wavelength scale at the rim of the chart. Distances of more than a half wavelength go full circle on the chart.

Example 10.6 ━━━━━━━━━━━━━━━━━━━━━━━━━━━━━━━━━━━━

Use the Smith Chart to find the input impedance of a 0.35 wavelength piece of 75 ohm cable that has a 150 ohm resistor connected as a load.

Solution:

First, the two impedances must be normalized by dividing each by the line's 75Ω characteristic impedance:

$$\text{The line impedance becomes } \frac{75}{75} = 1.0$$

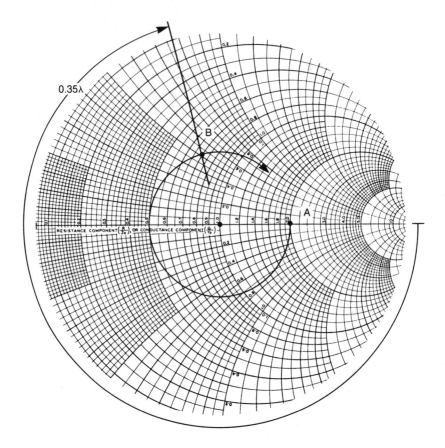

0.35λ

B

A

RESISTANCE COMPONENT(R/Z₀), OR CONDUCTANCE COMPONENT(G/Y₀)

Figure 10.16

The load impedance becomes $\dfrac{150}{75} = 2.0$

A dot is placed on the chart at 2.0 on the purely resistive axis (point A). This represents the normalized load impedance.

With a compass, a circle is drawn clockwise starting at point A and using 1.0, at the exact middle of the chart, as the center. Using the outside wavelength marks, a distance of 0.35 wavelengths is followed around the circle clockwise (toward the source) to point B. The normalized impedance here is:

$$Z_n = 0.68 + j0.48$$

The actual "real-life" impedance is found by multiplying both components by 75Ω.

$$Z_{in} = 75 \ (0.68 + j0.48)$$
$$= 51 + j36$$

10.12 PULSES ON LINES

A further insight into transmission lines can be obtained by looking at the behavior of pulses on transmission lines. This situation is typical of the address and data bus lines in microprocessor circuits.

Figure 10.17 shows a narrow pulse that starts out from a signal generator as a 10 volt amplitude. As it enters the line, it sees only the characteristic impedance of the line (50Ω) and so is reduced to a 5 volt amplitude. The other 5 volts are lost in the generator internal resistance.

After a certain amount of traveling time has elapsed, the leading edge of the 5 volt pulse arrives at the load end. To its amazement, it discovers that the

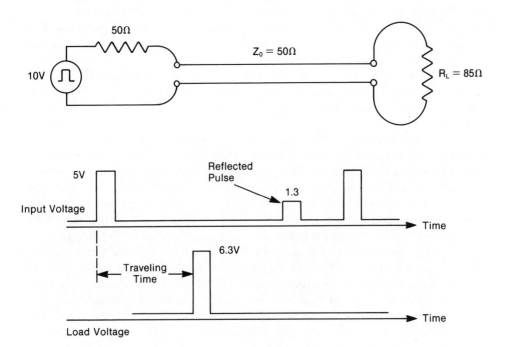

Figure 10.17 Pulses travel along a transmission line and reflect off the load mismatch.

load resistor has a different value from the line's impedance (85Ω vs. 50Ω). The load, therefore, cannot accept all of the power coming down the line and so, regretfully, some has to turn back. The mismatch at the load end results in a reflection coefficient of:

$$\Gamma = \frac{85 - 50}{85 + 50} = 0.259$$

This also gives us the ratio of the reverse voltage wave to the forward wave. The reverse voltage is then:

$$V_{\text{reverse}} = 0.259 \times 5.0$$
$$= 1.295 \text{ volts}$$

Because the load resistance is higher than the line's characteristic impedance, the phase of the reflected wave is the same as the forward wave at the point of reflection. Therefore, the waves add at that point and the total load voltage will appear higher than the pulse amplitude that is traveling down the line. This is similar to the situation described with filters, where the output voltage could be higher than the input voltage.

Again, after a second traveling time interval, the reflected pulse appears back at the pulse generator. Its energy is completely absorbed here because, in this example, the generator's internal resistance was chosen to match that of the line. If the resistances didn't match, there would be a second reflection and a weak "echo" would head back toward the load. Then it would, in turn, create a reflection of its own.

In a practical situation, the pulses will always need a certain rise time to get from zero up to full amplitude. Unless the mismatch is extremely bad, the reflections will not cause much of a problem as long as the traveling time is less than the rise time.

Narrow pulses can be used for testing transmission lines. By watching the input of the line as a series of slow, narrow pulses are sent into it, any breaks or abrupt impedance changes caused by pinches or sharp bends can be both identified and located. The location is easy if the cable's velocity is known. The time delay between the main pulse going into the line and any echo coming back will represent the round-trip distance to the fault and back. A commercial instrument manufactured for this purpose is known as a *time delay reflectometer*.

Another important point that can be learned from this discussion concerns the bad situation set up in Figure 10.18. In this case, a technician has connected an oscilloscope to a high-speed logic circuit. A coaxial cable has been used instead of a proper 10:1 probe. The input resistance of most oscilloscopes is 1.0 Megohms and this will be a bad mismatch for the 75Ω cable used. The output pulses from the first logic gate will be loaded, initially with the 75Ω impedance of the cable, not the 1 Megohm resistance of the scope. However, this isn't quite as serious as the reflection problem. If the output pulse from the first

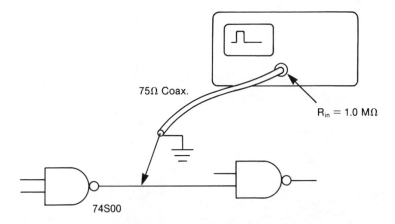

75Ω Coax.

$R_{in} = 1.0\ M\Omega$

74S00

Figure 10.18 A bad situation. A coaxial cable is used instead of a proper attenuating probe. The reflected pulse could create problems with the circuit being tested.

gate only stays Hi for a short time and then returns Lo, a second pulse with about the same amplitude could suddenly appear slightly later in time due to the reflection at the scope's input terminals. The result is a logic circuit that now has twice as many pulses going into the second gate as it should have. The moral of this story is to use a proper oscilloscope probe especially when fast rise-time pulses are being measured. The added resistance in series with the cable absorbs the return pulse.

10.13 CABLE LOSSES

When standard cables are used, the manufacturer will often have charts available that describe the cable losses for different lengths and different frequencies. A typical example is shown in Figure 10.19. For coaxial cables such as those included in the graph, a small portion of the loss may be due to signal leakage out through the shield. Another small portion will be due to losses in the dielectric insulating material. But, the main loss comes from the skin-effect resistance of the thin center conductor. The outer conductor has vastly more surface area and its resistive losses will be negligible in comparison.

The center conductor losses can be reduced by increasing the diameter of the wire. But, this would cause a change in the characteristic impedance of the cable unless the outer conductor diameter were also increased to maintain the diameter ratio. Therefore, larger diameter coax generally has a lower loss than thinner cable.

Figure 10.19 may seem a bit strange. The reason is that the vertical scale is calibrated in decibels and spaced logarithmically. The reason this was done was to produce straight lines for each cable's losses. Figure 10.20 contains essentially the same information but in a form similar to that used when filters were described. Now the seriousness of the losses at higher frequencies and for longer cable lengths becomes apparent.

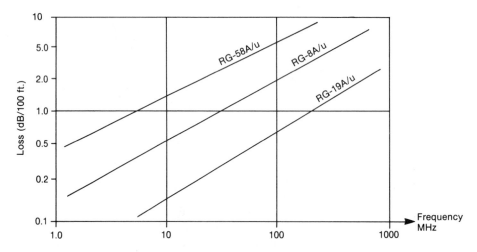

Figure 10.19 Total losses of standard coaxial cable at different frequencies.

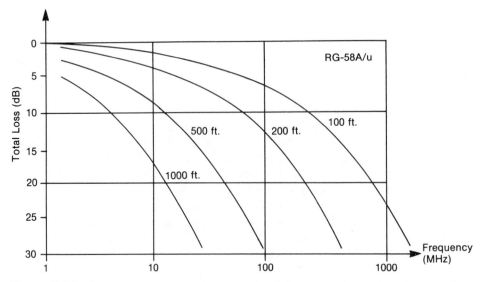

Figure 10.20 Coaxial cable loss vs. frequency. This graph presents the same information contained in Figure 10.19.

Table 10.2 Commercial transmission line characteristics.

Type	Impedance	Velocity Factor	100 MHz Loss dB/100 ft	Dimensions
Twin lead (TV)	300Ω	73%	1.4 dB	1.83 × 10.2 mm
RG 58A/U	50Ω	66%	5.3 dB	4.95 mm diam.
RG 59B/U	75Ω	66%	3.5 dB	6.15 mm diam.
RG 8A/U	52Ω	66%	2.0 dB	10.3 mm diam.
RG 19A/U	52Ω	66%	0.69 dB	28.8 mm diam.
Twin lead (#20 AWG)	72Ω	70%	7.7 dB	1.91 × 3.25 mm
Twin lead (#13 AWG)	72Ω	67%	3.8 dB	4.22 × 7.0 mm

10.14 NOISE PICK-UP

If transmission lines are not totally shielded, they are likely to pick up unwanted signals from random noise sources or signals on adjacent wires. In most cases, this is undesirable and techniques have to be examined to reduce the amount of pick-up.

Figure 10.21 describes a transmission line where one conductor is connected to ground. The impedance from each individual wire to ground is, therefore, not equal and this makes it an unbalanced transmission line.

As stray signals pass by the two wires, the ungrounded one is more likely to pick up signals from the "air" than the grounded one because of this impedance unbalance. The actual amount picked up depends on the impedance from the noise source to the wire and from the wire to ground. A voltage divider calculation will yield the amount picked up by each wire if the noise source voltage is known. If unbalanced lines are to be used, they should be

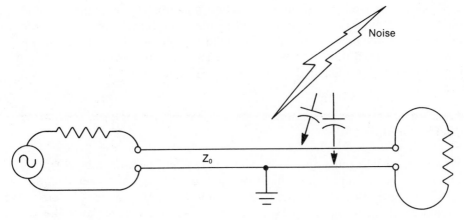

Figure 10.21 An unbalanced transmission line is very susceptible to stray noise pick-up.

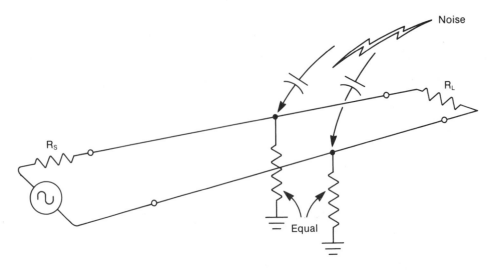

Figure 10.22 Balanced wires pick up equal noise voltages so that there is no resulting difference between them.

shielded as well as possible and their impedance kept as low as possible. Figure 10.22 describes a better way.

The balanced pair tries to carefully maintain an equal impedance to ground for each wire. This way both wires should pick up the same amount of noise and so there will be no difference between them. The wanted information being sent over the wires is carried as a voltage difference between the wires and any common voltage between the wires and ground is ignored (common mode rejection).

The subscriber loops that carry telephone signals to individual homes use a 600 ohm balanced pair to minimize noise pick-up. The 300 ohm "twin-lead" for television antennas is also a balanced pair. The term *twisted pair* refers to two wires twisted together partly to keep them together for constant impedance but also to ensure that both wires get an equal chance to pick up unwanted noise for optimum balanced operation.

10.15 CROSS TALK

When sets of wires are run close together they will pick up signals from each other. This is a result of the capacitive and inductive coupling between the sets of wires. Figure 10.23 shows the two coupling paths. The capacitive component creates a voltage in the second wire that travels toward each end of the second wire with the same polarity. The inductive component creates a different effect.

The induced voltage will be in series with the second wire and so the signal traveling toward the one end will be opposite in polarity to that traveling toward the other end.

To distinguish between the two ends, a set of standard names are used. The end of the second wire that is close to the source end of the first wire is known as the *near end*. At the opposite end of the cable is the *far end* which is close to the load of the first wire.

The net effect of this coupling is different at the two ends. At the near end, the two signals have the same polarity and so will add. The waveform at the near end should, therefore, be the same shape as the forward traveling signal on the first line. Notice that this might appear to be the "wrong" end in that the coupled signal is traveling back toward the source end.

At the far end, the two components have an opposite polarity so there will be partial or maybe even total cancelation. If the cancelation is only partial, the polarity of the small signal that survives at the far end could be the same as the original signal or inverted—depending on whether the capacitive or the inductive coupling was stronger.

Most of the time this coupling is a nuisance. The worst situation occurs when the second set of wires is carrying a signal in the reverse direction from the first set. Then the "listening end" is at the near end where the cross talk is the strongest.

However, this coupling also makes it possible to build *directional couplers.* This is the device that can be used to measure a sample of the wave flowing in

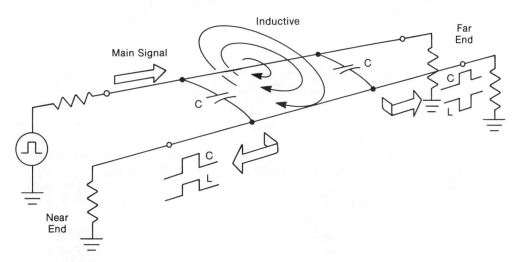

Figure 10.23 Cross talk is the result of capacitive and inductive coupling between two wires. The voltage capacitively coupled will be the same at both ends. The inductive component will create opposite voltages at the two ends.

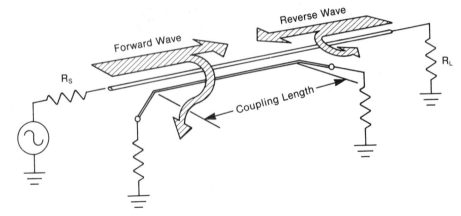

Figure 10.24 Directional coupler has two output connections. One will produce an attenuated sample of the forward signal flowing on the main line. The other gives a sample of the reverse.

the one direction on a transmission line independent of any reflected signals flowing in the reverse. The operation of a coupler is shown in Figure 10.24.

The length of parallel wiring necessary to pick up a signal depends on the rise time of the signal. For a given wire spacing, the maximum signal will be picked up when the traveling time over the pickup length is about one-half the rise time of the original signal. The strength of this maximum depends on the spacing between the two wires and not on any extra length.

10.16 HOW SHORT IS A TRANSMISSION LINE?

We started the chapter by saying that all lines really should be considered as transmission lines but, as we have seen, the calculations can be a bit messy. Whenever possible, it would be nice to ignore the lines and just use simple formulas such as Ohm's law. How short must a line be before we can ignore it? Part of this has already been answered. For pulses, the line's characteristics can generally be ignored if the traveling time on the line is less than the rise time of the pulses traveling on it.

For continuous sine wave operation we can be more precise. Consider the following condition as shown in Figure 10.25. A 20 volt, 500 ohm source is connected to a 500 ohm load through a pair of wires of various length. If the wires have zero length, we don't have to worry about transmission lines at all and the load power would be 0.2 watts—a case of optimum power transfer.

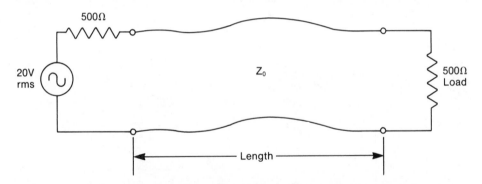

Figure 10.25 A line of variable characteristic impedance and variable length is used to connect a source to a matching load. The resulting power losses caused by the variations in impedance and length are shown in Figure 10.26.

Now, if we start with a line that has a characteristic impedance of 250 ohms, we have a VSWR of 2.0. The input impedance of the line can be calculated for various lengths and then the power that enters the line from the source can be calculated. Finally, these power values can be compared to the optimum condition of 0.2 watts and any loss determined. For example, for a 250 ohm line at a length of 0.15 wavelengths, the input impedance would be 170 + $j120$ ohms. The power entering the line and finding its way to the load would be 0.183 watts. This represents a power loss of 1.34 dB.

The results of these calculations are shown in Figure 10.26, along with similar calculations for a 100 ohm line (VSWR = 5.0) and a 50 ohm line (VSWR = 10.0). Each curve peaks at one-quarter wavelength since this is the worst impedance transformation. The losses all return to zero at one-half wavelength because the input impedance will be the same as the load impedance. The patterns repeat every half wavelength.

10.17 WAVEGUIDES

An alternative to the more conventional transmission line at microwave frequencies is a hollow, rectangular (sometimes circular or elliptical) metal pipe called a *waveguide*. Its most obvious difference from a coaxial cable is that it is totally empty—there is no center conductor. In fact, the main advantage of the waveguide is due to the absence of this conductor.

Waveguides will generally have a lower attenuation than coaxial cable and they are capable of handling much higher powers. They can also be purchased in a semiflexible form and so are about as easy to install as large diameter coax.

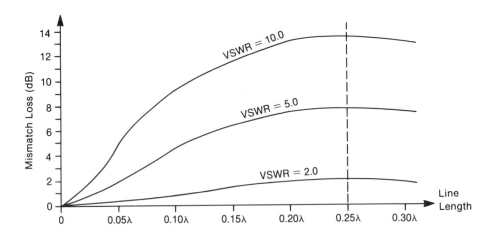

Figure 10.26 With equal source and load resistance, the power loss caused by using lines of differing length and characteristic impedance can be determined.

Unlike more conventional lines, the cross-sectional dimensions of waveguide are frequency-dependent. The inside width (the larger or "a" dimension) must be at least one-half wavelength at the operating frequency. This means that guides become larger as the frequency drops and, although it would be possible to use waveguide down to very low frequencies, it becomes impractical. Cost and mechanical factors limit the lowest practical frequency for the use of waveguides to about 2000 Mhz (2.0 GHz).

A waveguide of a given size is usable only over a 1.5 : 1 frequency range. Therefore, waveguides are not as general purpose as coax and other lines are.

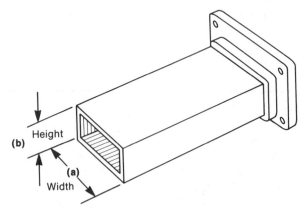

Figure 10.27 Rectangular waveguide with a connecting flange. The critical dimension is the width ("a" dimension). The height "b" determines the power-handling capabilities.

They are normally used where their lower losses and higher power handling are important.

When discussing more conventional lines, we described the electromagnetic wave moving down the wires and interacting with the voltages between the wires and the currents through them. Roughly the same thing happens in a waveguide. Here the electromagnetic wave bounces from side to side down the guide, setting up currents on the inner surfaces and voltages between the tops and bottoms of the guide.

The bouncing wave interferes with itself and creates very intricate patterns of electromagnetic fields called *modes*. Although a large number of mode patterns are possible, we will be dealing only with the one called the *TE 10 mode* (transverse electric field mode one, zero). The currents flow on the inside surfaces of the metal walls of the guide. Because of the very thin skin depth at microwave frequencies, no signal penetrates through to the outside surface of the metal. It is perfectly safe to touch the outside even when megawatts of power are flowing on the inside.

The electromagnetic field starts out in the guide with its electric field parallel to the side walls. But, it cannot proceed straight down the waveguide as it does with coaxial cable. If it tried to, the metal side walls would short out the electric field portion and the whole wave would die. Instead, the wave must start at an angle to the center line of the guide and proceed by bouncing diagonally off the side walls. With two components bouncing off opposite walls at the same time, an interference pattern results so that the total electric field is

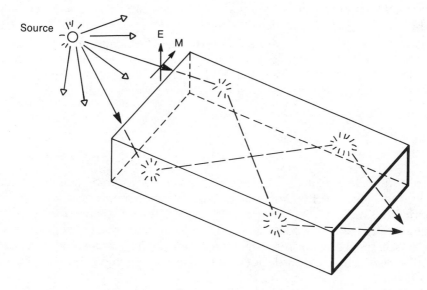

Figure 10.28 Two waves bouncing diagonally down the guide form interference patterns called *modes*.

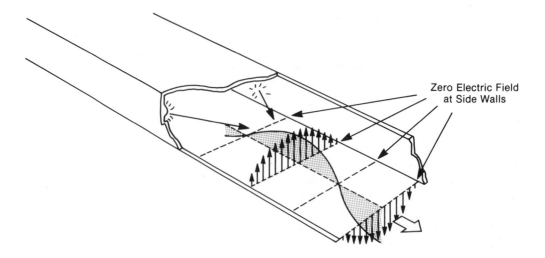

Zero Electric Field
at Side Walls

Figure 10.29 The TE-10 mode has one-half cycle of variation across the width of the guide and zero variation in strength across the height.

strongest at the center of the guide and zero at the side walls. The resulting mode pattern is shown in Figure 10.29.

How does the wave know what exact angle it is supposed to pick? It doesn't. Waves start initially into the guide at many different angles. Most die out but feed their power to the one that continues to survive—the one that happened to be at the correct angle of bounce.

Because of the bouncing back and forth, the wavelength measured along the center line of the guide is longer than it would be in free space. How much longer depends on the exact frequency and the guide's internal width. The relationship between guide wavelength and free space wavelength is:

$$\lambda_g = \frac{\lambda_0}{\sqrt{1 - (\lambda_0/2a)^2}} \tag{10.16}$$

This tends to make it look as if the signal is moving down the guide faster than the speed of light, which is impossible. The reason is simply that you are not measuring the wave in the direction it is moving. Imagine waves at the beach crashing straight onto the shore. If you were to measure the distance between wavepeaks in the direction from which they were coming, you would measure a true wavelength. But, if you decided to measure diagonally, you would get a longer distance.

Looking back at this formula, it shows that as frequency decreases the guide wavelength gets longer very rapidly as the angle of bounce increases.

When the free space wavelength is equal to twice the inside width ("a" dimension) of the guide, the waves will be bouncing back and forth off the side walls at right angles. The result will be no forward motion, the guide wavelength will be infinitely long, and the attenuation will also be infinite. This is called the *lower cutoff frequency* of the waveguide.

The attenuation curve for a sample piece of waveguide is shown in Figure 10.30. The lower cutoff frequency is at 3.0 GHz and the attenuation at this point is infinite because there is simply no forward motion down the guide. At slightly higher frequencies, the attenuation is dropping but is still high. There is now some forward motion but a large amount of bouncing is involved and normal resistive losses are involved at each bounce. Slightly higher frequencies give faster forward motion, less bouncing, and less attenuation. Eventually, another problem becomes dominant. At the higher frequencies, the skin depth is becoming thinner and so the resistance of the metal is increasing. Now, even though there are fewer bounces, the loss at each bounce is becoming higher. For very low losses, the inside surface of the guide is often silver plated.

The figure also shows that a second mode pattern could exist starting at 6.0 GHz. To avoid the complications of having two modes existing in the guide at the same time, the upper useful frequency limit is set to slightly less than this frequency. Waveguides are, therefore, operated over a frequency range that is less than 2:1.

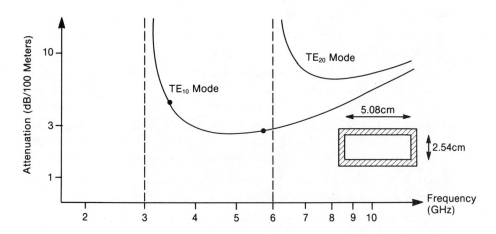

Figure 10.30 Attenuation curve for a sample of rectangular waveguide. The useful frequency range is from 3.5 to 5.5 GHz.

10.18 REVIEW QUESTIONS

1. Calculate the velocity of an electromagnetic wave in a material that has a dielectric constant of 2.80.

2. What is the wavelength of a 950 kHz signal in air and on a transmission line with PVC insulation?

3. Find the characteristic impedance of a transmission line if a sample length has the following characteristics: Capacitance = 208 pF, total Inductance = 3.1 μH.

4. What is the velocity of electromagnetic waves on a cable if a sample length has the following characteristics: Capacitance = 640 pF, total Inductance = 9.2 μH.

5. Two # 22 wires are spaced 2.0 mm apart in the air. What is the characteristic impedance of the line formed?

6. A coaxial cable uses polyethylene dielectric. The outer conductor has a diameter of 1.1 cm and the inner conductor 3.0 mm. What is the cable's characteristic impedance?

7. A 75 ohm cable is terminated with an 85 ohm load.

   a. What is the reflection coeffecient?

   b. What is the line VSWR?

   c. If a 35 volt sine wave were applied to the input end of this cable what would the highest and lowest voltages along the line be?

   d. What will the highest and lowest impedances be on the line?

8. What is the input impedance at 27.0 MHz for a 3.5 meter length of 50 Ω polyethelene coaxial cable that is terminated with a 62 ohm resistor?

9. Will there be any difference in the standing wave pattern of two identical lines, both of which have the same mismatch at their load end, but at the source end one is properly terminated and the other isn't?

10. What must the length (in meters) and characteristic impedance be for a "matching section" used to properly match a 100 ohm cable to a 200 ohm load at 3.8 MHz? All transmission lines are air dielectric.

11. On a normalized Smith chart, find the input impedance of a 50 ohm cable for the following conditions:

    a. Load = 75 ohms; length = 0.25 wavelengths

    b. Load = 75 ohms; length = 0.35 wavelengths

    c. Load = 60 + j20 ohms; length = 0.20 wavelengths

12. A time domain reflectometer is used to look for a fault in a long cable that uses PVC insulation. A reflected pulse appears 35 microseconds after one is sent into the cable. How far down the cable is the fault?

13. If the reflected pulse has the opposite polarity from the original transmitted pulse, what does this tell you about the fault?

14. Using Figure 10.26, what approximate mismatch loss will occur if a 0.20 wavelength piece of 50 ohm line is used to connect a 75 ohm source to a 75 ohm load?

15. What would happen if something like 5 meters of 50 ohm coaxial cable were used to connect a 75 ohm source to a 75 ohm load and the signals being carried extended over a very wide bandwidth?

16. What is the wavelength inside a waveguide for a 5000 MHz signal if the guide width is 3.3 cm? How does this compare to the free space wavelength?

17. What is the lower cutoff frequency of a waveguide with an internal width of 3.3 cm?

11. DATA COMMUNICATIONS

Data communications is a very broad topic. It generally refers to the transmission of information in a digital form, i.e., in a sequence of 0s and 1s. The original information may be digital to start with, such as the output from a computer; or it may be analog information such as telephone voice signals that have to be "digitized" first.

A common element in most aspects of data communications is that the communication is taking place between machines, not between humans. This makes the big difference between traditional analog and the newer digital communications. With humans involved, a very powerful processor exists (your brain, believe it or not) that can make perfect sense out of a very noisy phone message or watch a snowy, rolling TV picture for hours.

Electronic circuits, on the other hand, are extremely dumb by comparison. Not even the most powerful mainframe computer is capable of making the associations and inferences and listening for subtle changes in voice patterns that you can while listening to a friend on the phone. Think of what happens if, suddenly, you answer a phone call from a stranger with an extreme accent. Initially, the communications may be difficult, but slowly you adapt to his or her style and fewer mistakes are made.

This is why the topic of data communications is so involved. Every aspect of noise, distortion, modulation, and bandwidths, etc., must be carefully studied to give those poor dumb pieces of hardware the best chance of handling signals with a minimum of error. It also means that special coding techniques and error checks must be added to the system.

11.1 THE DIGITAL CHANNEL

The *channel* is simply whatever path the signal takes from the source to the destination. It could be a pair of short wires with virtually unlimited bandwidth and excellent signal-to-noise ratio. Or, it could be a pair of long wires with serious losses at higher frequencies, due to skin effect, and with a high noise pickup. Or, it could be a telephone channel with no DC capability and a 3 kHz (300

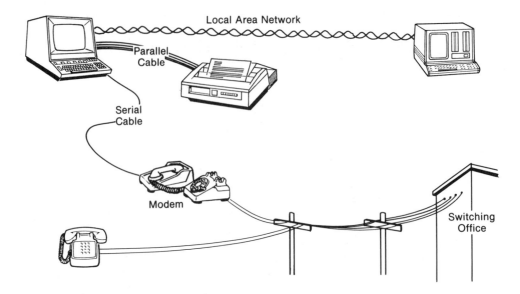

Figure 11.1 Data communications could involve short runs to computer printers or longer runs over telephone channels. The telephone system itself could be carrying analog voice, digitized voice, or computer data.

Hz to 3.3 kHz) bandwidth. A collection of these possibilities is shown in Figure 11.1.

A simple digital link is shown in Figure 11.2. Here a sequence of 0s and 1s from some source disappears into a cloud that represents long wires, losses, delays, and noise. In short, it represents a real-life "channel." Coming out of the cloud, the signal looks much worse. Now a simple circuit must decide when the input wave is high and when it is low. Filtering restricts the bandwidth to the essential minimum. Then a "slicer" circuit, consisting of a biased comparator, simply decides if the input is higher or lower than its bias point. Finally, a flip-flop is used to reset the timing. Its output should resemble the original data stream but the chances of mistakes are obvious. A small amount of noise likely wouldn't cause many Hi/Lo errors but could cause timing shifts. Larger amounts of noise start to show more frequent errors.

To improve the situation, attempts can be made both to clean up the channel before the Hi/Lo decision is made and to examine the sequence of 0s and 1s after the detector to see if the overall message or pattern makes sense. The after-detection methods involve the use of error correction codes. The before-detection methods are all those analog details described in previous chapters.

When a designer is faced with the initial task of creating a data communications system, there are a number of major choices that must be made:

- How many bits must be sent per second?
- How many bits will be carried at one time?
- Is the timing synchronous or asynchronous?
- Is the channel one way or two way?
- Is error checking needed?
- What happens if there is an error?
- Is modulation needed? If so, will it be AM, FM, PM, or a combination?
- What voltages and currents are to be used to represent the logic levels?

These are the major questions and they will be examined one at a time to see the advantages and pitfalls of each. There are numerous other questions that would likely also have to be answered as a design evolves.

If the data communications application is totally under the control of the designer, the design characteristics can be freely picked to best suit the conditions. However, if the design must connect to products from another company then the question of standards comes up. The major groups involved in setting these standards are:

Institute of Electrical and Electronic Engineers (IEEE)

American National Standards Institute (ANSI)

International Electrotechnical Commission (IEC)

Electronics Industries Association (EIA)

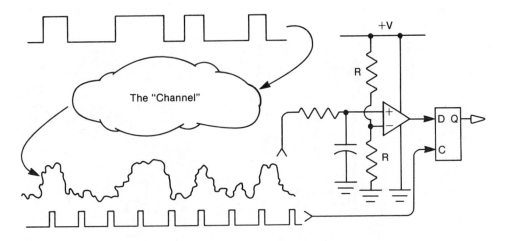

Figure 11.2 A noisy channel makes the Hi/Lo decision more difficult.

International Consultative Committee for Telephone and Telegraph (CCITT), a group of manufacturers and users

International Standards Organization (ISO), a United Nations Organization

11.2 BITS PER SECOND

Figure 11.2 showed a number of bits following one after the other over some sort of channel. The speed at which these individual bits can be sent determines the total time it will take to send a full message. Normally, it is desirable to keep this total time short so a large number of bits must be sent each second. As an example, the standard 25 by 80 character screen on most professional computers will display a total of 2000 characters. At a rate of 110 bits per second, which is a standard slow data rate, it will take 3 minutes and 20 seconds to fill the screen. This is painfully slow. At 9600 bits per second, the time will be reduced to a much nicer 2.3 seconds.

While high data rates are very convenient, they also mean wide bandwidths over the channels used. For a short pair of wires, this isn't much of a problem; but, for long distances or for access to the telephone network, bandwidth costs money.

What bandwidths are involved? This depends on the shape of the waveform used to carry the data. It is tempting to suggest square waves as the natural shape. However, square waves have an infinite number of harmonics and this adds up to a wide bandwidth as shown in Figure 11.3.

A square wave is simply a sequence of alternate 0s and 1s, i.e., 0101010101010101. Each cycle contains two bits. A 1000 Hz square wave is, therefore, sending data (monotonous data) at the rate of 2000 bits per second.

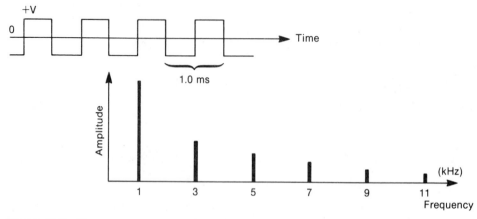

Figure 11.3 Frequency spectrum of a 1000 Hz square wave.

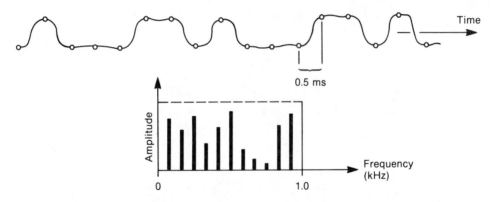

Figure 11.4 Random data, with "sine wave" shaping to reduce bandwidth, will have frequency components from DC up to half of the data rate.

However, 1000 Hz is just the fundamental frequency. A square wave also has all those harmonics that must be included if the shape is to be maintained.

When bandwidth is a problem, the square wave can be filtered to remove all harmonics and leave only the fundamental. The results are shown in Figure 11.4. The Hi and Lo data points appear to be connected by half sine wave segments. This is the minimum bandwidth signal. A random pattern of bits will have frequency components all the way from zero frequency (DC) up to half the data rate. Therefore, a data rate of 2000 bits per second using segments of sine waves will require a bandwidth of 0 to 1000 Hz.

Standard data rates are 110, 300, 600, 1200, 2400, 4800, 9600, 19200 bits per second. Of course, much higher ones are also available and a designer could pick any nonstandard value if he or she felt it absolutely necessary.

11.3 SERIAL VS. PARALLEL

If we have a pair of wires that have a bandwidth of 600 Hz, for example, the fastest we can send data is 1200 bits per second. To increase the speed, we could add a second pair so that two bits are being carried at a time—in parallel. If eight individual wires are used, the total capacity climbs to 9600 bits per second.

A parallel path can reduce the need for high data rates on any one of the wires but there are problems. As the distances become great, so does the cost of the parallel wires. There will also be a problem of keeping the timing of the pulses the same on all wires. Therefore, parallel wires are normally not used for more than about 10 meters. Printers that are located close to computers very often use a parallel cable to save costs. Another excellent example is the IEEE-

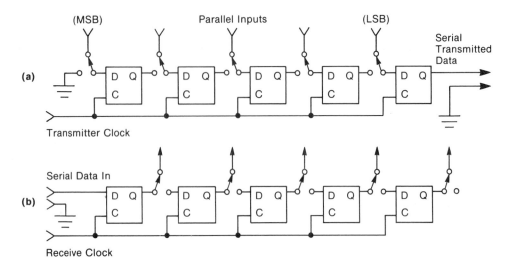

Figure 11.5 Shift registers act as parallel-to-serial (a) and serial-to-parallel (b) converters.

488 instrumentation bus used to connect measuring instruments to computers. This will be discussed later in this chapter.

Computers usually handle data in parallel form. If a serial channel is to be used to carry information to a distant point, some sort of parallel-to-serial converter will be needed at the sending end. At the receiving end, the opposite will be needed—a serial-to-parallel converter. These two converters are shown in Figure 11.5. The basic element in each is a *shift register*, which is simply a set of flip-flops connected end to end. At the transmitting end, the flip-flops are all loaded at the same time (parallel loading); each has its own input. Then the internal connections are changed and the data is shifted to the right from one flip-flop to the next. The flip-flop at the right end is connected to the serial path so data enters the line least significant bit first.

Parallel paths don't really need multiple wires. If only one bit at a time is allowed on one wire pair, multiple wires will be needed. But, if one wire is allowed to carry two or three bits at a time, fewer wires are needed. Figure 11.6 shows how multiple bits are carried. One bit requires only two voltage levels, two bits need four voltage levels, and three need eight levels. There is no theoretical limit to the number of bits and voltage levels that can be used but, assuming a maximum voltage that can be used, more levels mean a smaller spacing between levels. With noise and other distortion, the levels will become harder to distinguish and the advantage of digital transmission over analog will be lost. A "digital" path with an infinite number of levels is exactly the same as an analog channel.

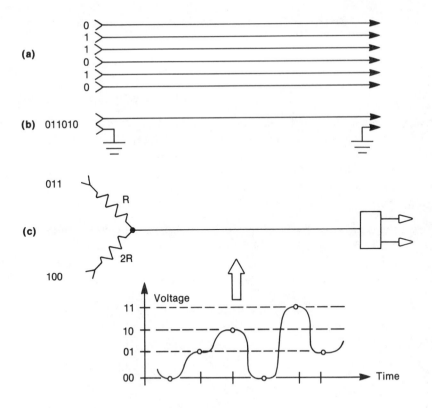

Figure 11.6 Parallel (a), serial (b), and multilevel (c) channels.

11.4 BITS PER SECOND AND BAUD RATE

With a serial channel carrying binary (two-level) data at a certain rate, there is a simple relationship between the bandwidth and the data rate:

$$\text{Bandwidth (Hz)} = 1/2 \times \text{Data Rate (bps)} \qquad (11.1)$$

With parallel wires, each carrying one bit at a time, the same relationship holds true as long as the data rate on any one wire is used.

To accommodate more than two levels on any one pair of wires, a new term is introduced—*baud rate*. This describes how fast the line is changing without worrying about how many levels are involved. For binary (two-level) transmission, where only one bit at a time is being carried, the bit rate and the baud rate are the same.

$$\text{Baud rate (no units)} = \frac{\text{Total Data Rate (bps)}}{\text{Number of Simultaneous Bits}} \qquad (11.2)$$

$$\text{Bandwidth (Hz)} = \frac{1}{2} \times \text{baud rate} \qquad (11.3)$$

For example, if a line carries two bits at a time and the total data rate is 4800 bits per second (bps), the baud rate would be 2400 and the required bandwidth would be 0→1200 Hz (assuming sine wave shaping). The line would require four voltage levels to represent the two bits being carried simultaneously (see Figure 11.6 (c)).

11.5 SIMPLEX, DUPLEX AND HALF-DUPLEX

It is a rare situation when data is transmitted in one direction only, with no messages expected back. Even for a simple printer, a reverse channel is normally used to indicate ready/not ready (out of paper, broken ribbon, busy printing, etc.). In the rare case that no reverse channel is used, the situation is known as *simplex*.

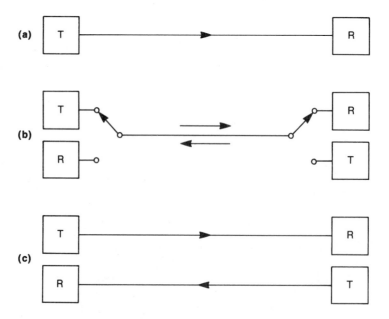

Figure 11.7 Three possibilities exist for the reverse channel: no channel (a), time switching (b), and permanent (c).

If one channel is alternately used for the forward and then the reverse channel, we have a *half-duplex channel*. The 8- or 16-bit data bus on a micro-computer circuit board is an example of a half-duplex, parallel bus. It is bidirectional, allowing the processor to write to memory and peripherals and also to read data from the same places. The reading and writing cannot, of course, occur at the same time nor is direction arbitrary.

This example can be used to illustrate a few other points that concern data communications. The microprocessor, in this case, is the "boss." It tells all other circuits on the bus (a set of parallel lines) whether it is writing (transmitting) or reading (receiving). If two "bosses" were to be used (a direct memory access (DMA) controller added, for example), the problem of *contention* arises where two or more circuits might try to transmit at once. With two bosses, a *protocol* must be set up so that the two bosses can agree on who is in charge at any one time.

When full communications are needed in both directions at the same time, two channels are used; this is a *full-duplex* operation. The forward and reverse channels might have their own separate wires or they might share the same wire pair and use modulation to carry their data at different frequencies. The forward and reverse channels don't need to have the same capacity; their data rates can be individually set to suit the information being carried.

11.6 SYNCHRONOUS AND ASYNCHRONOUS

When a stream of data enters a receiver, there must be some sort of timing signal present that indicates when each individual bit is present. It isn't good enough just to know that data is coming in at, say, 300 bits per second. There must also be a definite phase relation between this data and the sampling clock. An example of this was shown back in Figure 11.2. There are two ways to maintain this timing synchronization:

1. *Synchronous data transmission* carries the clock signal along with it. Thousands of bits can be sent in a continuous stream without having to stop for timing checks and synchronization. The circuitry needed at both the sending and the receiving ends is a bit more elaborate but is readily available in integrated circuit form. We will see in a moment how this clock signal is included. Synchronous transmissions are used mainly for higher data rates where this extra complexity can be justified.

2. *Asynchronous data transmission* does not include a continuous timing signal with the transmitted signal. At the receiving end, an oscillator is set to within a few percent of the correct data rate and then a synchronizing signal is sent about every 7 or 8 data bits to set the correct phase

relationship. Between synchronizing bits, it is assumed that the receiver clock won't wander out of phase by more than a half of a bit interval. The advantage of this is simplicity—a very necessary feature many years ago when mechanical clutches and electric motors were used to control timing. Today, integrated circuits make the task simple.

The disadvantage of asynchronous transmission is the continuous waste of channel time every 7 or 8 bits to reset the timing. Because of this, asynchronous communications is limited to the slower data rates where this loss of efficiency isn't critical.

Figure 11.8 compares the signals used for the two transmission methods. The asynchronous signal begins with what is known as a *start bit*. The channel switches from the Hi to the Lo state and holds this for 1 bit interval. This signals the receiver that a 7- or 8-bit character is to follow. These character bits follow one at a time, least significant bit first. When all bits have been sent, the channel returns to the Hi condition and holds this for a minimum of 1 or 2 bit intervals. This serves as a *stop bit(s)* and guarantees that the next Hi-to-Lo start bit will be recognized. With 1 start bit, 8 data bits and 1 stop bit, for example, asynchronous transmissions are only 80% efficient.

The synchronous signal begins with a unique pattern of 8 or 16 bits to set the initial synchronizing of the receivers clock both for individual bit timing and also for the group of 7 or 8 bits that make up a character. Once synchronized, this pattern won't have to be repeated for hundreds or even thousands of bits.

The clock signal must be carried along with the data stream. Figure 11.9 and Table 11.1 show some ways in which this can and cannot be done.

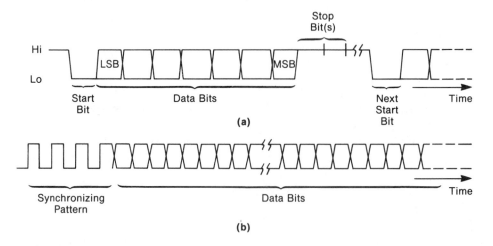

Figure 11.8 Asynchronous (a) and synchronous (b) data transmission.

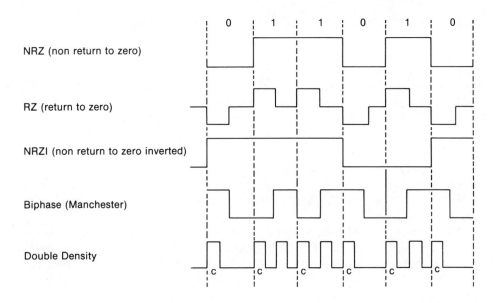

Figure 11.9 Five commonly used data waveforms. Some carry the bit clock signal along with them and so are self-clocking.

The standard pattern of Hi's and Lo's is known as a *non return to zero* (NRZ) waveform. It and a related signal, NRZI, do not carry their clock with them. A receiving system could use a phase-locked loop to lock onto the highest frequency of these waveforms and then double it to obtain the bit clock. The problem is that there could be long periods of time where strings of 1s and 0s

Table 11.1 Each waveform has its own format for carrying the 0 and 1 data bits. This results in wide ranges of bandwidths.

Type	*Format*	*Self-Clocking?*	*Bandwidth*
NRZ	0 = low voltage 1 = high voltage	No	$0 \rightarrow 0.5 \times$ bps
RZ	0 = negative pulse 1 = positive pulse	Yes	$0.25 \rightarrow 1.0$
NRZI	0 = change to opposite 1 = no level change	No	$0 \rightarrow 0.5$
Biphase	0 = change from Hi to Lo 1 = change from Lo to Hi	Yes	$0.5 \rightarrow 1.0$
Double D.	0 = no extra pulse 1 = extra pulse between fixed clock pulses	Yes	$0.5 \rightarrow 2.0$

would appear by chance. The PLL would not have anything to synchronize to in this case so mistakes would start to appear.

With NRZ and NRZI, the transmitter and receiver must be a little more intelligent. The transmitter could be designed to watch for five Hi's or five Lo's in sequence and then insert a dummy bit of the opposite value. The receiver would watch for the same sequence and then remove the next bit. This is known as *bit stuffing*. As long as the time constants in the PLL can last for a 6-bit interval, there will be enough variations in the waveform for the loop to stay locked.

The *return to zero* (RZ) and *double density* waveforms obviously carry their clock pulses with them and so the circuitry at both the transmitter and the receiver needn't be as elaborate. Their disadvantage is their higher bandwidth requirements.

The *biphase* or *Manchester II* encoding is a good compromise. It has a level transition of one direction or the other right in the middle of every bit interval. This can easily be used for timing purposes by the receiver. Manchester encoding is used by the ETHERNET local area network which is a very high speed (10 megabits per second) data path between computers.

11.7 BASEBAND VS. MODULATION

For many applications, the binary voltage levels can be applied directly to a set of wires and carried to a receiver that simply has to separate one level from another. The frequency range will usually cover DC to some high frequency that depends on the data rate. This is a *baseband* system.

However, there are times when it is necessary to have the data modulate a carrier so that the entire signal can exist at some higher frequency. With modulation, there is no DC component to the signal and the result is a *broadband* system. Modulation allows several channels to be carried on one wire either in the same or different (duplex) directions. Modulation can also be used to spread the information over a wider bandwidth to gain some degree of noise immunity.

The telephone network is one example where modulation must be used. The telephone line, once it passes through the switching system, cannot handle DC signals. DC levels between the home phone and the switching office are used to indicate that the receiver has been lifted "off the hook" and also for pulse dialing. The frequency range available for voice or data is limited to 300 Hz at the low end and something like 3000 or 3300 Hz at the upper end.

Another reason to use modulation involves a problem with long wires at lower frequencies. As we saw in the chapter on transmission lines, characteristic impedance has an important effect on the way signals pass down a line. At low frequencies, this impedance varies slowly from a pure resistance caused by DC losses, to a different and more consistent value determined by capacitance, inductance, and skin effect resistance at higher frequencies. For some applications, it is desirable to move to a higher frequency to achieve this consistent impedance even if it involves slightly higher losses. Remember that the bandwidth between 0 and 5 kHz is the same as 45 to 50 kHz but the ratio of upper to lower frequency (percentage bandwidth) is much smaller at the higher frequency so performance over this bandwidth will be more consistent. All frequency components will travel with virtually the same speed and attenuation.

Broadband transmissions can use any of the modulation systems described in the previous chapters. In fact, some applications use two modulation types simultaneously. The three basic modulation types—AM, FM, and PM—are shown along with a representative data signal in Figure 11.10.

A modulated channel requires a modulator at the sending end and a corresponding demodulator at the receiving end. If a reverse channel is to be used—and it very often is—it will need a modulator at its sending end and a demodulator at the other end. The result is a *duplex channel*. Figure 11.11 shows the required components and assumes that a separate pair of wires is used for each direction. However, the extra pair of wires used for the reverse channel

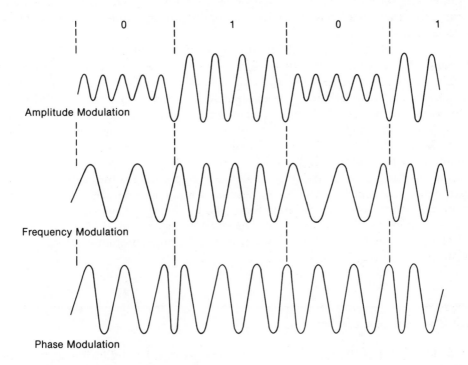

Figure 11.10 Any of the three basic modulation types can be used to modulate a carrier with data.

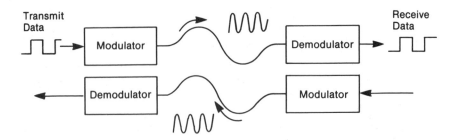

Figure 11.11   Modulators and demodulators are needed at both ends of a broadband duplex channel.

aren't really needed. If modulation is already being used, it is a simple (?) matter to assign one frequency range to the forward channel and a different range to the reverse channel. Then the full-duplex operation can be carried on the one pair of wires. The reason the question mark was placed after the word *simple* is that there must now be some agreement between the users at each end concerning who is going to transmit on what frequency. And, agreement is never simple. The solution is to make the frequency bands interchangeable by either manually or automatically switching the modulator and demodulator frequencies around.

The combination of a MOdulator and DEModulator and the frequency separating devices (duplexer) necessary to carry the two directions on one wire pair is known as a *modem*. Figure 11.12 shows the frequency spectrum and the block diagram of a Bell 103-type modem. This is a standard model used over the telephone system to simultaneously send and receive asynchronous data (with start and stop bits) at 300 bits per second. Frequency modulation (FM) is used. Within the low band, a 0 is transmitted as a 1070 Hz tone and a 1 as a 1270 Hz tone. In the high band, the 0 is 2025 Hz and the 1 is 2225 Hz. The required frequency range needed to include sidebands totals 400 Hz for each band. The two bands are kept below 2600 Hz because some telephone channels use this frequency for a separate signaling tone. The modem used to originate the call will transmit on the Lo band and receive on the Hi band. The answering modem will detect the Lo band tone coming in and automatically switch its transmitter to the Hi band.

11.8   FM MODULATOR

The design of the FM modulator itself is interesting. The following section describes the method used for the transmitter section of the Motorola MC6860 integrated circuit.

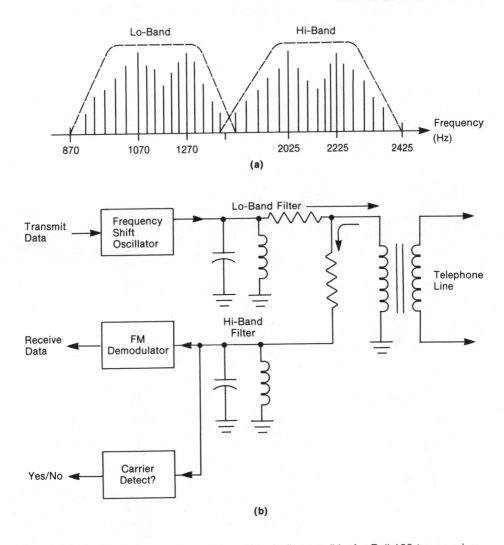

Figure 11.12 Frequency spectrum (a) and block diagram (b) of a Bell 103-type modem. This model handles one forward and one reverse channel of asynchronous data at 300 bits per second each.

It is not an easy task to build a sine wave oscillator that must shift between the low band frequencies of 1070 and 1270 Hz, especially if it must be done in integrated circuit form without any inductors. But, it is easy to build a square wave oscillator to operate over the same frequency range. The problem is to get rid of the harmonics. A square wave, as we have already seen, consists of a fundamental and third, fifth, seventh, etc., harmonics. For this application,

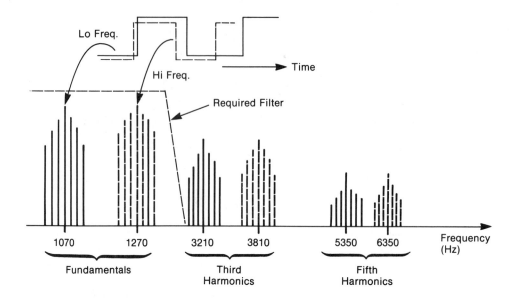

Figure 11.13 An FM modulator based on square waves would need elaborate filters to remove unwanted harmonics.

the harmonics must be filtered out. The problem is that the filter must be fairly elaborate to give a steep enough slope to pass all desired sidebands below 1370 Hz and yet get rid of unwanted harmonics that begin at 2910 Hz.

The MC6860 modem "chip" uses a variable frequency square wave oscillator at a higher frequency. It runs a shift register that is connected to a digital-to-analog converter. The result is shown in Figure 11.14. The output waveform now contains a strong fundamental and much weaker harmonics. The strongest harmonics are 15 and 16 times the fundamental frequency and so are easily filtered out.

11.9 HIGH-SPEED MODEMS

To transmit information at a rate of 9600 bits per second over a telephone channel, several bits must be carried at one time. The reason is that the channel bandwidth is only 2700 Hz (300 to 3000 Hz). If 4 bits are carried simultaneously, the baud rate would drop to 2400 and the required data bandwidth would be 1200 Hz. The 4 bits would require 16 different modulation states. By selecting a carrier signal in the center of the telephone channel and modulating it, the re-

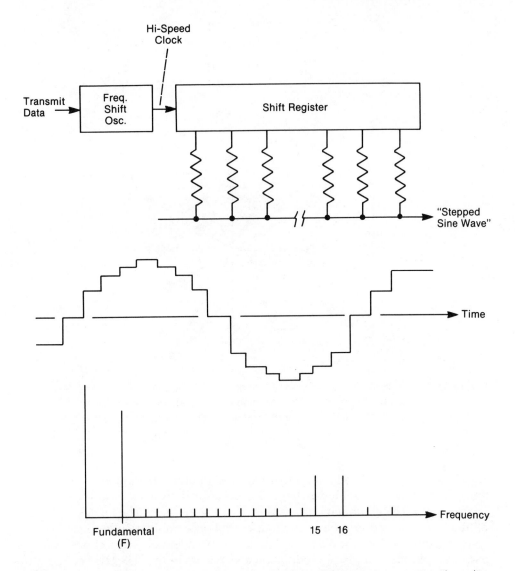

Hi-Speed
Clock

Transmit
Data

Freq.
Shift
Osc.

Shift Register

"Stepped
Sine Wave"

Time

Fundamental
(F)

15   16

Frequency

Figure 11.14 Digitally synthesized "sine wave" has a low harmonic content and can be
easily filtered to provide a true sine wave.

sulting sidebands will extend at least 1200 Hz on either side and so the channel
will be completely filled. But, at 9600 bps there is no room left for a reverse
channel. An example of a standard 9600 bps modem is the Bell-type 209A.

For best error performance, a modulation system must be selected to give
each of the 16 states the best chance of being detected and not being confused
with a neighboring state. The Bell 209A modem uses a combination of both

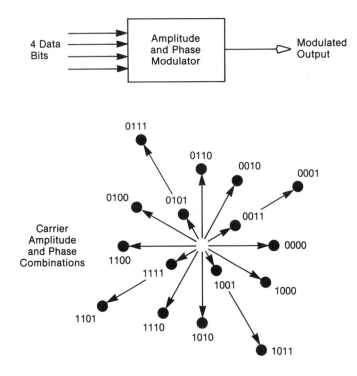

Figure 11.15 At 9600 bits per second, 4 bits must be carried at one time. The Bell 209A modem uses a combination of amplitude and phase modulation to provide the best chance of distinguishing one 4-bit combination from the next.

amplitude and phase modulation to obtain this (quadrature amplitude modulation). The modem output can have 3 different amplitudes and 12 different phase angles. The combinations are shown in Figure 11.15. The angles and amplitudes are selected in such a way that each 4-bit combination is spaced equidistant from the surrounding combinations. (Incidentally, this simultaneous use of amplitude and phase modulation also occurs in the 3.58 MHz color subcarrier for television broadcast.)

11.10 ERRORS

Regardless of the transmission system used, there is a chance that errors in the transmitted data will occur. Why do they occur, what happens if they do occur, and what can be done to reduce the possibility?

11.10.1 Why Do Errors Occur?

Errors occur when one or more bit values are read wrong. A 1 may be accidentally interpreted as a 0, for example. The reasons can be any of those described in earlier chapters. Noise is a common problem. It may be a continuous, random noise or it may occur in bursts when nearby motors or electrical switches are operated. Noise could also be the result of intermodulation in a nonlinear circuit and it could also be the result of unwanted signal pickup from adjacent wires.

Errors can also occur because of phase and delay distortion. This is a common problem that arises when information is sent through narrow bandwidth channels (such as the phone system). The simple filters used will likely create large delay variations near the band edges. If the data transmitted crowds near these edges in an attempt to gain extra bandwidth, excessive phase shift may result in errors. Figure 11.16 indicates the typical delay variations that occur in a voice-grade telephone channel. Later we will see that this can be corrected by using "delay equalizers."

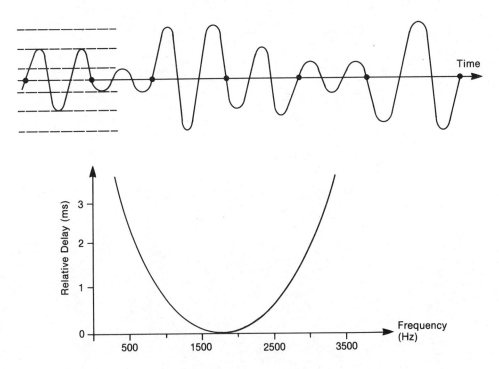

Figure 11.16 Typical delay variation over the bandwidth of a voice-grade telephone channel. This variation could cause errors to occur if attempts are made to use the full bandwidth.

Another source of errors is frequency shift. Within the telephone system, a lot of modulation and demodulation takes place in order to place multiple telephone calls on single-trunk cables. As a result, the "voice" frequency at the receive end could be off by as much as 10 Hz. With one person talking to another, this won't matter but with high-speed data it could. For cases where the user supplies his or her own wires without any extra frequency multiplexing, this problem won't exist.

11.10.2 How Are Errors Detected?

This depends on who wants to know. Normally, the people involved with setting up a communications channel will do an initial check to see how good it is. From an analog point of view, the "eye pattern check" can provide a good "feel" for the quality of the channel. This is shown in Figure 11.17.

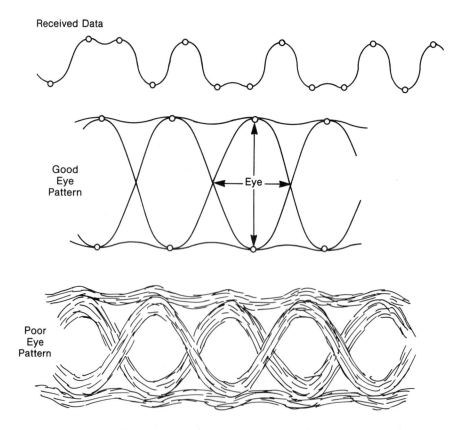

Figure 11.17 An "eye pattern" displayed on an oscilloscope will provide a good indication of the overall channel quality.

At the receiving end of a channel, and before any further processing takes place, an oscilloscope is connected to the line. It is adjusted to provide a periodic sweep at some submultiple of the data rate. The image will, therefore, be a composite of the random Hi and Lo voltages being received. Ideally, all amplitudes will be the same and the 0 crossovers will occur at well-defined points. Extra noise will cause the pattern to blur in both the horizontal and vertical directions and the "eye" will become smaller. Delay distortion will cause a horizontal widening of the lines and a resulting horizontal closing of the eye.

A more quantitative test can be made if a full-duplex channel is available. If the received data can be "looped back" to the sending end, it can be compared with the transmitted data. Test equipment can then be used to count the total number of bits sent and the total number of wrong bits. Error rates can now be measured in "wrong bits per million."

Now that the people involved in creating and using the "channel" have satisfied their curiosity, it is far more important for the data itself to know that errors have occurred. This involves inserting extra bits into the transmitted data so that some automatic checking and comparing can be made. This does slow

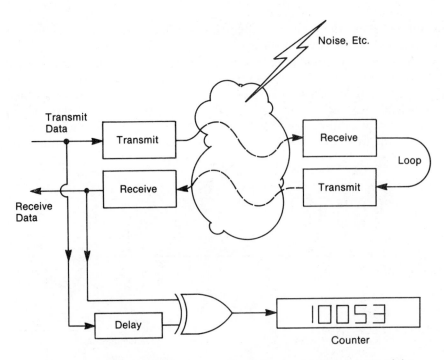

Figure 11.18 Looped-back data can be used to determine total error rates of the forward and reverse channel. The output of the Exclusive-OR gate is Hi only if its two inputs are different.

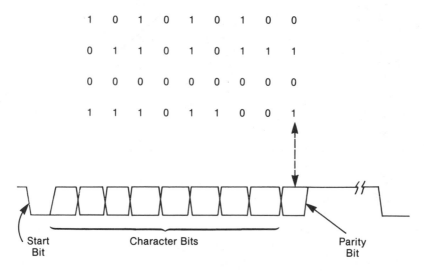

1 0 1 0 1 0 1 0 0

0 1 1 0 1 0 1 1 1

0 0 0 0 0 0 0 0 0

1 1 1 0 1 1 0 0 1

Start Bit

Character Bits

Parity Bit

Figure 11.19 Parity bits can be added to detect single-bit errors within a character grouping. This example uses even parity.

down the effective data rate by a small amount but there is no sense in sending data at a high speed if half of it is wrong.

The simplest method of error checking is to add a *parity bit* after each 7 or 8 bits (one character). The people involved must initially agree on whether *odd* or *even* parity is to be used. For example, with even parity the extra bit will automatically be made a 0 or 1 so that the total of the data bits plus the parity bit will be an even number of 1s. At the receiving end, a continuous check can then be made to see that the received pattern always has an even number of 1s.

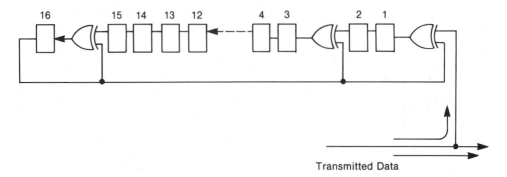

Transmitted Data

Figure 11.20 Cyclic redundancy checks use long-shift registers with internal feedback to accumulate a unique pattern based on all the bits in a long block of data.

If one bit within the character grouping is wrong, the error will be detected and the receiver can request a retransmission. However, if an even number of bits are wrong, the error will not be detected.

More elaborate error checks involve the use of *cyclic redundancy checks* (CRC). This involves the use of special 12- or 16-bit shift registers at both the sending and receiving end. The data at each end is copied into these shift registers where it continues to circulate for a considerable length of time. Periodically, the register at the sending end and the receiving end can be compared to see if they hold the same pattern. If they do, no errors have occurred within the past block and transmission can continue. If the patterns don't match, the shift registers must be reset and the previous block retransmitted. The benefits of the CRC method of error detection are that groups of wrong bits can be detected and the channel doesn't need to be burdened with the continuous sending of parity bits. The CRC pattern is only 16 bits or less and it need only be sent at the end of each block of several thousand bits.

11.10.3 How Are Error Rates Reduced?

The first step is to find out why they are occurring. Is it a noise problem, a stray signal being picked up, or is it due to delay or frequency shift in the channel? The source can usually be determined with analog measurements and "loopback" tests.

Once the cause is found the solution may be to use better cables, possibly with shielding, or to locate them further away from sources of noise such as motors and electrical controls. The solution may also be to use balanced lines instead of unbalanced lines and to be more careful of characteristic impedance and terminations.

If the telephone system is being used, it is possible to lease a permanent line to one location instead of always having to redial the same number and having to suffer with a different path each time. With leased lines and an extra monthly payment, it is also possible to have improvements made to the channel in both noise and delay characteristics. These are specified by the phone company under *line conditioning*. It is also possible to add your own delay equalizers. Figure 11.21 shows a circuit that will adjust the delay over a narrow bandwidth. The amount of the delay is determined by the loaded Q of the resonant circuit and the center of the added delay is determined by the resonant frequency. If several of these are used, the variation across a channel can be reduced. The total delay will increase but that won't matter.

When all else fails, slow down. Error rates will decrease as data rates are reduced. The higher rates are useless if the data being transmitted is not correct.

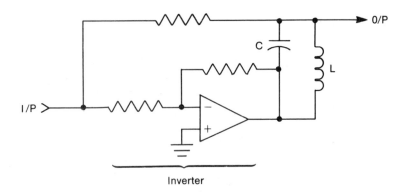

Inverter

Figure 11.21 Delay equalizer circuit adds a time delay that depends on the resonant circuit Q. The delay acts over a narrow bandwidth.

11.11 CHANNEL CAPACITY

For a particular channel with a limited bandwidth and a specific signal-to-noise ratio, what is the maximum number of bits per second that can be carried before the error rate becomes excessive?

If the line is to carry only two voltage levels (binary signals) in a baseband mode and the noise level isn't too high, the answer is easy. The maximum data rate will be limited to twice the bandwidth of the channel (Equation 11.1). This assumes that the channel isn't noisy enough to cause a 0 to be accidentally interpreted as a 1 or vice versa, and that delay and frequency shift problems aren't severe.

If multiple levels (i.e., 4, 8, 16, . . .) are to be used, the capacity climbs and the calculation becomes a bit more difficult. Assuming that the channel has a maximum signal amplitude that it can handle, then as more levels are used they must be spaced closer together. This closer spacing will increase the probability that random and other sources of noise will cause errors to occur. Signal-to-noise ratio, therefore, plays an important part in channel data rates and error rates.

If, for example, two bits were carried at a time over a channel and the four voltage levels to be used were 0, 1.0, 2.0, and 3.0 volts, we would have the situation shown in Figure 11.22. At the receiving end of such a system, a group of circuits would have to make voltage decisions based on the "half-way" points between the individual transmitted levels. For example, a decision would have to be made as to whether the voltage is above or below 2.5 volts. If above, then it was likely originally sent as 3.0 volts. If below, then it was likely originally sent as 2.0 volts.

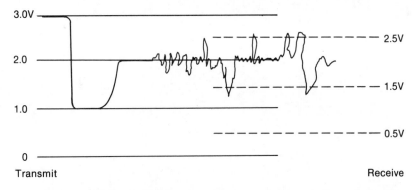

Figure 11.22 With 2 bits sent at one time, 4 voltage levels are needed. The receiver must make a decision half-way between these levels. Random noise of plus or minus 0.5 volts would cause errors.

Random voltages will make the individual levels wander around. As long as any particular level doesn't wander more than plus or minus 0.5 volts, no errors will occur. However, with random noise, any voltage could occur. The low voltages are most probable but there is a small chance that very large voltages could occur occasionally.

A random noise waveform and the probability of any particular voltage occurring is shown in Figure 11.23. The probability curve is symmetrical at about 0, indicating that positive and negative voltages are equally likely to occur. The curve is highest at the 0 point because the 0 voltage crossovers occur most frequently.

The probability of any voltage over a certain level occurring is given by the area under the curve and above the particular voltage. Areas at both ends of the curve are included because both positive and negative voltages will cause errors (different errors). The probability of voltages of either polarity and

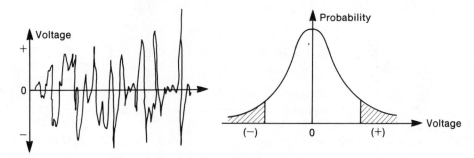

Figure 11.23 Typical random noise voltage (a) and the probability of any voltage occurring (b).

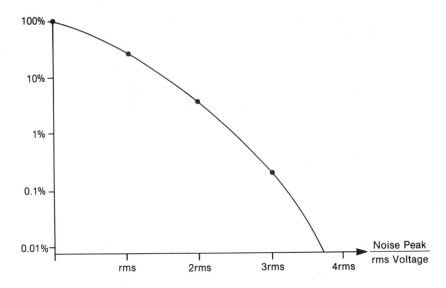

Figure 11.24 Probability of a random noise voltage with either polarity being higher than a specific value.

greater than a particular value occurring is given by Figure 11.24. For example, the probability of a voltage greater than the rms value of the noise occurring is 36%, and the probability of a voltage greater than three times the rms value occurring is 0.26%. This is the general shape of all error rate curves in data communications as long as the errors are due purely to random noise.

Figure 11.25 summarizes the error rates that will be experienced in channels using 2, 4, 8 and 16 levels when the signal-to-noise ratio is known. When the signal-to-noise ratio on a limited bandwidth channel is known, the absolute maximum data rate is given by a formula developed by Claude Shannon of Bell Labs in 1948. It is based on situations similar to that just described.

$$\text{Max. data rate (bps)} = \text{BW} \times \log_2 \left(1 + \frac{S}{N} \right) \qquad (11.4)$$

where BW is the channel bandwidth in Hz
S/N is the signal-to-noise ratio of the channel

This is a theoretical maximum based on random white noise only and assumes that some very elaborate error checking codes are used to reduce errors to an absolute minimum. In practice, it is next to impossible to achieve these maximums because other types of noise and distortion are present.

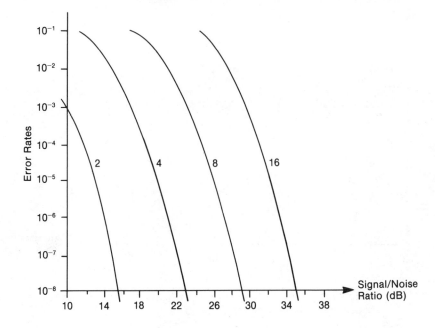

Figure 11.25 Error rates for channels carrying multilevel digital signals. The signal-to-noise ratio is measured in dB.

11.12 DATA COMMUNICATION STANDARDS

Because so many wires, connectors, voltages, waveforms, and pieces of software are involved in the applications of data communications, a number of standards have evolved to create some order. Some cover very small parts of the total topic while other standards are much more encompassing. Four of the popular standards are briefly introduced here.

11.12.1 RS-232C

This is a simple standard that describes a set of wires and connectors suitable for carrying baseband, full-duplex, serial data one bit at a time (two voltage levels). Three of the wires (transmit, receive, ground) are essential to the sending and receiving of the data. However, the standard also includes extra wires to carry control signals between the two pieces of connected equipment.

The standard is normally used for limited-distance communications between a computer and a printer, terminal, or modem. Not all of the wires are

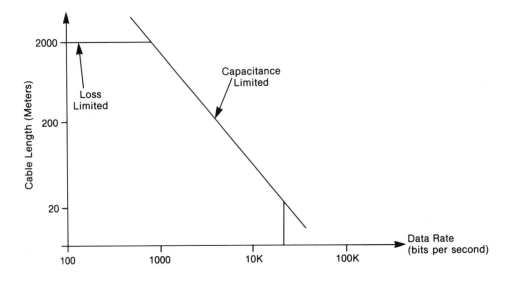

Figure 11.26 The maximum recommended cable length using the RS-232C standard depends on the data rate to be used.

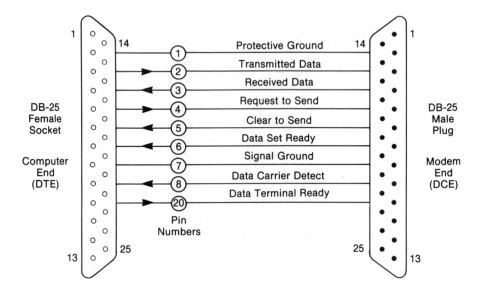

Figure 11.27 25-pin connectors and the major wires used for an RS-232C cable.

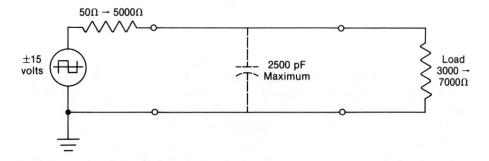

Figure 11.28 Each of the signal and control lines works in an unbalanced manner with one of the few ground wires. The load resistance must lie between 3000 and 7000 ohms. The source resistance can have a much wider range. The source voltage is negative to represent a 0 and positive to represent a 0 but the actual voltages can range from 5 to 25 volts.

needed for every application and many of the control lines are often left out. Unfortunately this means that this standard cannot always be relied upon as a standard.

The long wires in the cable are not treated as transmission lines and there is no attempt at matching the characteristic impedance. The load resistance is much higher than the cable's impedance and so the capacitance between wires has the dominant effect.

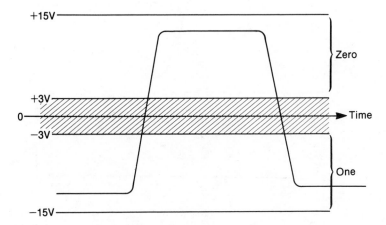

Figure 11.29 The receiver treats any voltage between +3 and +15 as a 0 and the opposite polarity as a 1. The speed of the transition is reduced to control cross talk between wires.

11.12.2 RS-422

For longer distance wires and greater data rates, the Electronics Industries Association (EIA) has produced a more recent standard. This one specifies the wires only. They work in pairs as balanced lines to minimize noise pickup and they are terminated with a resistance that matches their characteristic impedance. The line driver and the receivers are both designed for balanced operation. (RS-423 is a variation of this that uses an unbalanced driver.)

The driving voltages, measured relative to ground, are 6 volts. This results in a difference of 12 volts between the two wires. At the receiver, a difference of only 0.2 volts is needed. With one wire more positive than the other by 0.2 volts, the result is a 1. With the same wire more negative than the other, the result is a 0.

11.12.3 IEEE-488

This standard is also known as hp-IB and GPIB. It is an outgrowth of an instrumentation bus standard developed by Hewlett Packard. The bus carries data in an 8-bit parallel manner between high-speed devices such as computers and measuring instruments. It is also used as a local area network (LAN) to connect multiple personal computers together so that they can share information with each other. IEEE-488 covers a much greater number of the variables than do either of the previous two standards. Now, the wires, the connectors, and the actual signaling protocol are described in the standard.

The bus connects one controller and a number of talkers and listeners. It is the job of the controller to ensure that only one user is talking on the line at any one time, although anyone can listen. The individual wires and the devices

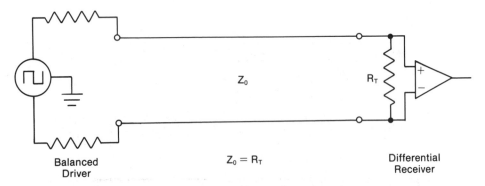

Figure 11.30 Balanced wires and proper attention to terminating impedance characterize the RS-422 standard.

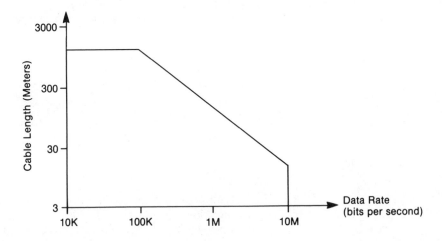

Figure 11.31 RS-422 data rates again depend on distance but both are much greater than for RS-232C.

that connect to them are shown in Figure 11.32. All lines carry standard TTL voltage levels but the Hi and Lo are reversed (ZERO >2.0 volts, ONE <0.8 volts).

The 3 handshaking lines allow an asynchronous transfer of each 8-bit character. This means that a listener must acknowledge receipt of each byte before the next is sent. The result is that devices of different speed capabilities can be mixed on the same bus. (The Motorola 68000 microprocessor uses a similar asynchronous data bus structure.) The five control wires are used to pass status signals back and forth to and from the controller to indicate that a device has some information it would like to send, etc. Each device has a five bit ad-

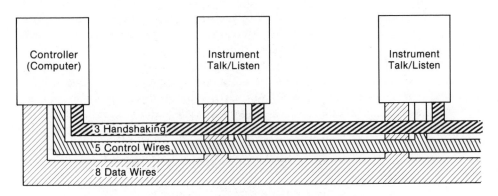

Figure 11.32 IEEE-488 bus wires consist of 8 data wires, 8 ground wires, 3 handshake wires, and 5 control wires.

dress code so that the controller can send or receive information from any of 31 specific devices.

Data rates depend on both the capabilities of the individual devices on the bus (the asynchronous handshaking structure looks after this) and also the overall length and total number of instruments connected. The maximum length is 20 meters and the highest data rate is one megabyte per second (remember that the bus handles a byte at a time).

11.12.4 ETHERNET

ETHERNET evolved from a 1972 research project at Xerox Corporation. It is a very high speed (10 Mega bits per second) local area network used to connect computers, disc drives, printers, etc., together. Unlike the network possibilities with the IEEE-488 standard, no single controller is used. Each device acts as a "peer" with the same privileges as any other device on the line. As such, ETHERNET includes a "line-access protocol" that arbitrates the problem of who is going to talk at any given time. There are two line-access protocols in general use for local area networks without controllers. We will take a brief look at the other method first and then look at the ETHERNET protocol.

At a business meeting with many people sitting around a table it is difficult to control who is talking. The result is that two or more are trying to speak at once and someone else is always trying to cut in. Some managers have resolved the problem by taking a tennis ball into the meeting and having it passed around from person to person. You can only talk when you are holding the ball. When you are done, you pass it to the next person. If they have nothing to say, they simply pass it to the next.

The electronic version of this is called a *token-passing network*. An appropriate electronic message is passed around the network from one device to the next. If a device has a message to send, it does so immediately after receiving the "token." If not, it passes the token to the next device and just listens. The result is a simple and very workable approach to network control but a device with a lot of information to send may find that it has to do a lot of waiting.

The ETHERNET approach is much wilder! Every one tries to talk at once—almost. The proper name for the technique is called *carrier-sense multiple-access with collision detection* (CSMA/CD). When a device has some information to send it first listens to the bus (a coaxial cable). If no one else is talking, it begins and it sends a "packet" of information. This, for example, could consist of 10 or so lines of text on the screen of a terminal with some extra bits added at the front to indicate who is sending the information and who it is intended for.

However, another station may do exactly the same thing. The network can be as long as 2.5 km and, because of the propagation delay along the cable, a distant station would not immediately know that someone else was using the

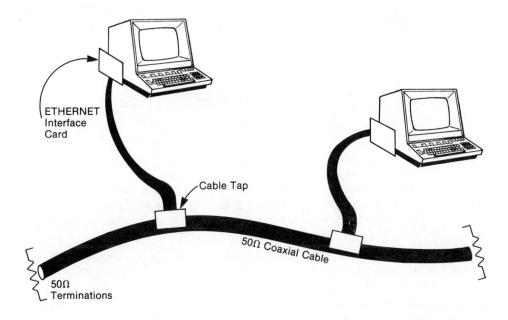

ETHERNET
Interface
Card

Cable Tap

50Ω Coaxial Cable

50Ω
Terminations

Figure 11.33 A local area network using ETHERNET connects high-speed devices together. The maximum cable length is 2.5 km.

shared channel. In this case a "packet collision" occurs and all information is lost. To solve this problem, any device that is sending also continues to monitor the cable and, when a collision is detected, immediately stops transmitting information and sends a brief "jamming" signal so no one else will get involved in the fight. Then the two devices involved each wait a random amount of time and try again. Because of the special brief jamming signal, any other device will wait an even longer amount of time so as not to confuse the issue. One of the original two will then try again. If a collision occurs on the next try, both stations immediately quit and wait an even longer but still random amount of time. When a packet successfully reaches its destination and is checked for errors (32-bit cyclic redundancy codes are used), the receiving station sends a brief acknowledging packet. The sender can then continue with its message if it needs to.

The packet structure is shown in Figure 11.34. It is sent directly on the 50 ohm coaxial cable (baseband) without a carrier, and uses Manchester encoding to include clock pulses. On the cable, a 1 is represented by a 0 volt level and a 0 by a minimum negative 2-volt level. The time for each bit is 100 nanoseconds so that the overall bit rate is 10 mega bits per second. The waveform is rounded to look like sine wave segments and to minimize excessive high-frequency content.

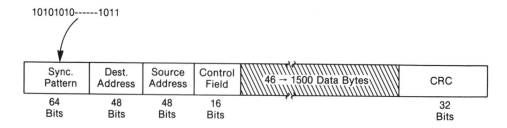

Figure 11.34 An ETHERNET packet consists of a 64-bit synchronizing pattern and a 48-bit destination address followed by a source address of the same length. Then a 16-bit control field is sent. The data itself can range from 46 up to 1500 bytes. Finally, a 32-bit error checking pattern is sent. The minimum spacing before the next packet can occur is 9.6 microseconds.

The access protocol to the shared cable sounds like a horribly inefficient operation. However, damaged packets will be quite short as all transmitters will immediately turn off when a collision is detected. The minimum size will be a function of the maximum length of the cable used. If most successful packets are near the longer limit, the ratio of success time to failure will be high. Measurements have shown successful data rates as high as 98% of the channel's maximum capacity.

11.13 REVIEW QUESTIONS

1. Why does the whole world appear to be "going digital"?

2. Give five examples of information being carried digitally.

3. What are the two most important characteristics of a "channel"?

4. Why are "standards" important?

5. A particular character code uses 9 bits for each character. Extra circuitry adds in an error-detecting parity bit. What data rate is needed for serial transmission of 50 characters per second? What bandwidth would be required for baseband transmission using only two voltage levels?

6. If a channel could handle two bits at a time, what bandwidth would be needed to send 100 characters per second? Each character contains 7 bits plus a parity bit plus one start and one stop bit.

7. Describe the channel requirements necessary to handle the same data at the same error rate as in question 6 but with three bits sent at a time.

8. For long paths, what are two disadvantages of parallel transmission of bits on separate wires?

9. What digital device converts parallel data into a stream of serial bits?

10. In terms of channel efficiency, what is the advantage of synchronous transmission over asychronous?

11. Why must modulation be used to send digital information over the telephone network?

12. Which transmission method uses modulation—baseband or broadband?

13. Why do some 300 bps modems have a switch on them that says ANSWER/ORIGINATE?

14. What is the advantage of continuously leasing a telephone line for data transmission instead of using the dial system?

15. Calculate the absolute maximum data rate (bps) that could be sent over a 10 kHz channel that has a signal-to-noise ratio of 20 dB.

16. If a computer is supposed to be sending a logic ONE over an RS-232C cable, what two pins on the DB-25 connector would you measure a voltage between? What magnitude and polarity would you expect this voltage to be?

17. What makes the RS-422 standard superior to RS-232C for long-distance, high-speed data transmission?

12. ANTENNAS AND PROPAGATION

Some areas of communications carry their signals continuously on lines and cables without ever letting them get loose into free space. However, when long distances, mobility, and a large number of listeners spread over a wide area are involved, it becomes very practical to send radio signals through the "air." Even for some "point-to-point" work, the losses associated with an antenna and radiated signal can be substantially less than for any transmission line. This chapter deals with the antennas used to launch these signals into open space and to pick them back up at the receiving end. It also describes how the signal travels through space and what factors affect it.

12.1 ANTENNA INTRODUCTION

An antenna is nothing more than an impedance-matching and directing device. It simply takes an electromagnetic wave that is traveling down a transmission line and lets it float off into free space. The presence or absence of air has nothing to do with the radiation. The antenna will normally also concentrate or beam the radiated energy in some general direction. An antenna is not a transducer; no change in energy form takes place as happens, for example, with a loudspeaker.

A transmitting antenna is no different from a receiving antenna other than for some possible changes to handle high transmitter power levels. However, as our description continues, we will find that it is sometimes helpful to think of the transmitting antenna as having "directing" properties and the receiving antenna as having "gathering" properties, the two being very closely related.

12.2 SMALL DIPOLE ANTENNAS

The simplest place to start is with two short pieces of wire placed end to end (without touching) and suspended up in the air with nothing else around. A balanced transmission line brings a signal up to the pieces of wire. If the two wires are extremely short, almost nothing happens. The electromagnetic signal coming up the cable sees an open circuit and so all the energy turns around, disappointed, and heads back. If the wires are a bit longer—for example, about 0.1λ end to end—things start to improve. The signal coming up the transmission line sees the extra length of wire and gets curious. At one particular instant, current flows out on the left wire and in on the right wire. This is the situation shown in Figure 12.1. A half-cycle later in time, the directions will reverse. As the currents flow, a magnetic field is set up around the wires and in the same direction for both wires.

But where does this current flow to? It looks as if there is an open circuit at the end of each wire. Because the wires are still close together, there will be some capacitance between them. The current that enters the one wire will loop back to the other wire through this small capacitance, flow through the second wire and back to the transmission line.

If there is a capacitance, there must be a voltage difference between the two pieces of wire and so an electric field must also exist. The results are similar to what was happening all along the transmission line before the signal got to these two wires. The big difference is that the two wires are not parallel as they were in the line but now point away from each other. The change in the total electromagnetic pattern around the wires is the key to what happens next.

At all places around the wires, we have an electric field at right angles to a magnetic field. This condition creates an electromagnetic wave that is capable of existing without benefit of current or voltage. This wave will move away

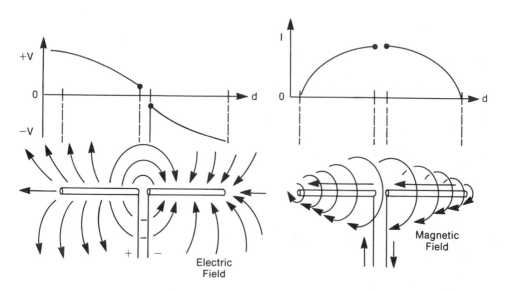

Figure 12.1 The simple dipole antenna with its electric and magnetic fields.

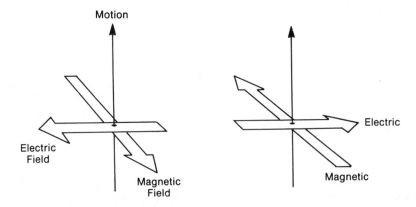

Figure 12.2 An electric and magnetic field crossing at right angles will move in a
direction at right angles to both.

from the wires in a direction at right angles to both the electric and magnetic
fields. Figure 12.2 shows this "vector" relationship and also shows that, when
both the electric and magnetic fields reverse direction a half-cycle later, the
direction of movement will still be the same.

Meanwhile, back on the transmission line. If the wires are extremely short,
almost all the energy coming up the line will be reflected back down the line. As
the short wires increase in length, a small amount of energy permanently
moves away from the region and so is not reflected. The bulk of the energy,
however, pauses temporarily in the electric and magnetic fields and then returns back down the line. Therefore, we have three fields around the wires.
There is an electric field and a magnetic field, each of which may extend a fair
distance out from the wires. But these are only temporary energy storage fields
and any energy not used for other purposes must return to the line. The third
field is the electromagnetic field that is capable of existing by itself. It forms
from parts of the electric and magnetic fields.

The energy on the transmission line sees the equivalent circuit shown in
Figure 12.3. This consists of an inductance that represents the separate magnetic field, a capacitance that represents the separate electric field, and a resistance that represents the energy that has combined and moved away from the
vicinity of the wires. This energy isn't lost and converted to heat as in a normal
resistor. It still exists as electromagnetic energy someplace out in space but it is
lost as far as the transmission line is concerned and isn't coming back. The
resistive component is known as the *radiation resistance*, to describe its origin.

For short wires, the capacitance dominates the inductance and the radiation resistance is very small. Most of the energy is, therefore, reflected back
down the cable. As the length of the wires increases, both the inductance and
the radiation resistance increase until there is a particular length where the re-

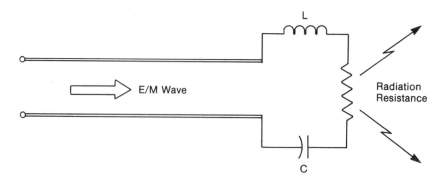

Figure 12.3 The transmission line sees a complex impedance at its end that represents the electric field (capacitance), the magnetic field (inductance), and the departing electromagnetic field (radiation resistance) of the antenna.

actances balance out and only a resistance is left. This variation of impedance with the length and also the diameter of the wires is shown in Figure 12.4.

The two parts to the impedance graph show that the radiation resistance gets steadily stronger as the overall length increases and the reactance decreases. This means that the Q is very high for short lengths and decreases as the element length increases.

The optimum length, from an impedance point of view, is just a little bit shorter than 0.50λ, depending on the wire diameter. The resistance at this point is about 72 ohms. This particular antenna length is commonly used simply because it provides a good match for a standard cable that has a characteristic impedance of 75 ohms. It should be understood that all lengths could act as antennas as long as they can be matched to their transmission lines. For very short lengths, this is difficult because the matching components could end up with losses that consume more power than the antenna radiates.

The impedances shown on the previous graphs represent fields that exist a considerable distance (several wavelengths) all around the antenna wires. If anything else is within this volume, the impedances would change somewhat.

Example 12.1 ═══════════════════════════════

What is the input impedance of a dipole that has an end-to-end length of 0.35λ and a diameter of 0.001λ? What are these dimensions at 50 MHz? What component would be needed to make the input impedance purely resistive? And what would happen if this component had a Q of only 50?

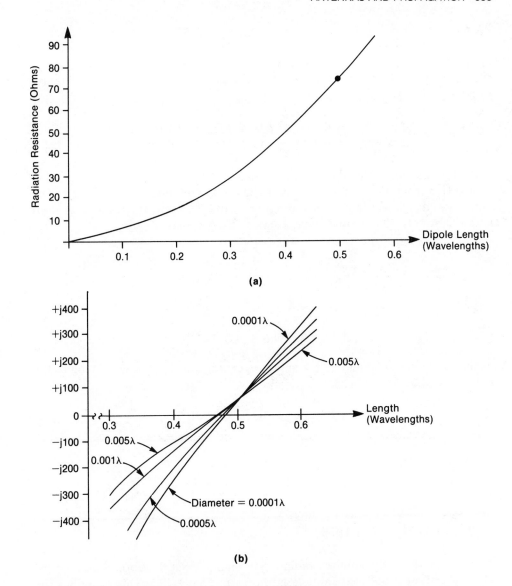

Figure 12.4 The equivalent impedance of the dipole antenna changes with overall length and also with element diameter. The resistive component is shown in (a) and the reactive component in (b).

Solution:

Figure 12.4 shows the input impedance to be about $37 - j230$ ohms. One wavelength at 50 Mhz is $300 \times 10^6/50 \times 10^6 = 6.0$ meters. The dimensions will therefore be:

$$\text{Length} = 0.35\lambda = 6.0 \times 0.35 = 2.1 \text{ meters}$$
$$\text{Diameter} = 0.001\lambda = 6.0 \times 0.001 = 6.0 \text{ mm}$$

To correct the input impedance, a reactance of $+j\,230$ ohms should be added in series. This reactance would be provided by a $0.73\mu H$ inductor at 50 MHz. The input impedance would now be a purely resistive 37 ohms. If the inductor used had a Q of only 50, it would have a series "loss" resistance of

$$R_s = \frac{X_L}{Q} = 230/50 = 4.6\Omega$$

This represents

$$100 \times {}^{4.6}/(4.6 + 37) = 11\%$$

of the total resistance and so the same percentage of the power flowing to the antenna will be lost as heat.

$$11\% \text{ power loss} = 0.51 \text{ dB}$$

12.3 RADIATION PATTERN

Now that we have the antenna radiating its signal, what direction does it go in? The pattern for a half-wave dipole is shown in Figure 12.5. This describes the relative strength of the electromagnetic wave in each direction. There is virtually no radiation directly off the ends of the wires and the strongest signal travels away at right angles to the wires. Viewed in three dimensions, the pattern would appear donut shaped.

Two numbers are used to describe radiation patterns—beamwidth and gain. *Beamwidth* is the total angle between the 3 dB points on the radiation pattern. For the half-wave dipole, this angle is 78°, indicating that the signal is sent out over a wide angle. *Gain* describes how the total energy that enters the antenna from the transmission line is concentrated in any one direction. If we

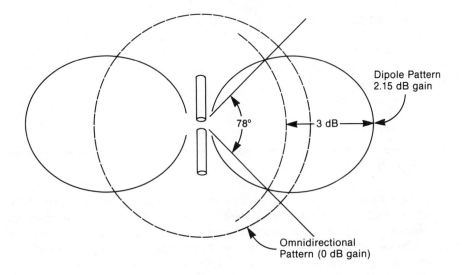

Figure 12.5 Radiation pattern for a half-wave dipole.

have a very small antenna that radiates equally well in all directions, it would be assigned an arbitrary gain of 0 dB and serve as a reference for all other patterns. Such an antenna is known as an *isotropic radiator*. All antennas that have some apparent "gain" do so by sacrificing radiation in some directions so that it can be concentrated in others. The half-wave dipole has a gain of 2.15 dB when measured broadside to the wire. It achieves its gain in this direction by sacrificing radiation out along the axis of the wires. (Figure 12.5).

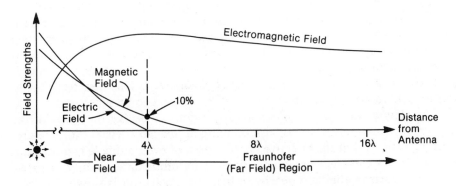

Figure 12.6 The electric and magnetic field strengths decrease rapidly. The electromagnetic field builds up within the other two and then decreases in strength 6.02 dB every time the distance is doubled.

A radiation pattern must be measured far away from the antenna. Otherwise, the measuring system will pick up the electric and magnetic fields that are close to the wires in addition to the electromagnetic (E/M) field. It is only the E/M field that survives to carry the signal over vast distances and so only it should be measured.

12.4   PROPAGATION AND RECEPTION

As you move away from an antenna in any one direction, the strength of the electromagnetic field will decrease at the rate of 6.02 dB every time the distance is doubled—the same as any sound or light signal from a point source decreases. The reason, in all cases, is simply that the signal is spreading out over and ever-widening area. None of it is actually getting lost.

It is interesting to compare this "attenuation rate" with that of a lossy transmission line. For example, the line might have a loss of 2.0 dB every 100 meters, whereas the antenna has a "loss" of 6.02 dB each time the distance is doubled. Notice the different specifications. If we arbitrarily say that at 100 meters from the antenna and 100 meters down the cable the signal strengths are identical, then 100 meters further along, the cable signal will be only 2.0 dB weaker and the signal from the antenna will be 6.02 dB weaker. The cable wins so far. At 400 meters, the cable loss will be 6.0 dB and the antenna "loss" 12.04 dB. At 800 meters, the cable loss is 14 dB and the antenna "loss" is 18.06 dB. But at 1600 meters, the cable loss is 30 dB and the antenna only 24.08 dB. From here on out, the antenna system will have the stronger signal. Take a look at the number of amplifier boxes hanging on the overhead wires for the cable TV system in your city and you will get a good idea of the relative attenuation rates of even high quality coax transmission line compared to the radiated signal.

The same type of antenna that is used to transmit a signal can also be used for its reception. All of the previous characteristics such as impedance, gain, and beamwidth still apply but it is helpful now to add one other. A receiving antenna can be thought of as a big fishing net up in the air catching the signal. The bigger the net, the more signal is captured. This new term is then *effective capture area*. It is directly related to the gain to an antenna.

The skinny wires of a half-wave dipole don't look like they would have much capture area but what you are thinking of instead is visible surface area. Figure 12.7 shows that the effective capture area of even a dipole is quite large. As it consumes the electromagnetic field in its immediate vicinity, it causes some other parts of the field nearby to bend in. The result is something like a vacuum cleaner. The effective capture area for a half-wave dipole is:

$$\text{Area} = 1.64\lambda^2/4\pi \text{ (sq. m.)} \tag{12.1}$$

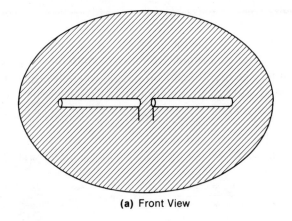

(a) Front View

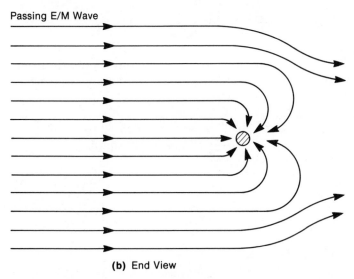

(b) End View

Figure 12.7 Effective capture area of a half-wave dipole.

For the isotropic "point source" antenna, the capture area is:

$$\text{Area} = \lambda^2/4\pi \text{ (sq. m.)} \tag{12.2}$$

Notice that the dipole area is 2.15 dB larger—the same as its "gain."

To realize this gain and use the capture area effectively, the receiving antenna must be oriented "broadside" to present the largest cross-section to the incoming signal and it also must be "polarized" so that the receiving wires lie in the same plane as the transmitting wires. If the two were oriented at right angles to each other, they would be "cross-polarized" and no signal would be received.

Example 12.2

Figure 12.8 presents a very specific problem. At a frequency of 100 MHz, a power of 1.0 watts is fed to a dipole antenna. A matching dipole is located 25 km away. Both antennas are aligned broadside to each other and "polarized" for best transmission and reception. How much power is picked up?

Solution:

The capture area of the receiving half-wave dipole at 100 MHz is:

$$\lambda = \frac{300 \times 10^6}{100 \times 10^6} = 3.0 \text{ meters}$$

$$\text{Area} = \frac{1.64\lambda^2}{4\pi} = \frac{1.64\,(3.0)^2}{4\pi} = 1.175 \text{ sq. m.}$$

At 25 km radius, if the signal came from an omnidirectional "isotropic radiator" it would have spread the 1.0 watts of energy over an area of (surface area of a sphere):

$$\text{Area of "ball"} = 4\pi r^2 = 4\pi(25 \times 10^3)^2$$
$$= 7.854 \times 10^9 \text{ sq. m.}$$

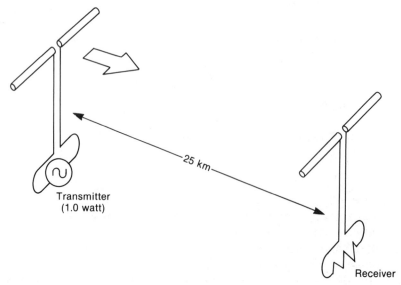

Figure 12.8 One dipole sends a signal 25 km to another. What is the total loss?

The "density" of the electromagnetic wave at this distance would be:

$$\frac{1.0 \text{ watt}}{7.854 \times 10^9 \text{ sq. m.}} = 1.273 \times 10^{-10} \text{ watts/sq. m.}$$

But the transmitting antenna isn't isotropic. It has a certain amount of gain because it isn't radiating equally well in all directions. At right angles to the wire's axis, the density will be 1.64 times stronger (2.15 dB gain). Therefore, at the receiving antenna, the density will be:

$$1.64 \times 1.273 \times 10^{-10} = 2.088 \times 10^{-10} \text{ watts/sq. m.}$$

With a capture area of 1.175 square meters, the power picked up by the receiving antenna and fed into its matched transmission line will be:

$$
\begin{aligned}
P_{received} &= 1.175 \times 2.088 \times 10^{-10} \\
&= 2.454 \times 10^{-10} \text{ watts} \\
&\text{or } 245.4 \text{ pico watts}
\end{aligned}
$$

(96.1 dB loss)

On a 75Ω transmission line, this would be 136 microvolts—quite a strong signal considering that a good 100MHz voice-bandwidth receiver can work very well with a 0.5 microvolt signal.

This whole procedure can be condensed into one formula:

$$P_{received} = \frac{P_T \; G_T \; G_R \; \lambda^2}{16 \; \pi^2 \; d^2} \text{ (watts)} \tag{12.3}$$

where P_T = transmitted power in watts
   P_R = received power in watts
   G_T = gain of transmitting antenna (not in dB)
   G_R = gain of receiving antenna (not in dB)
   λ = wavelength in meters
   d = distance between antennas in meters.

Occasionally, the strength of a radio signal at a certain location is given in volts per meter. This is the voltage between two points in the air one meter apart and oriented "broadside" and with the same polarization as the transmitting antenna. Theoretically, it should be possible to hold a voltmeter up in the air and measure this voltage. However, the results wouldn't be exactly right because the meter leads would also pick up a signal in addition to what was measured at the lead tips.

The electric field strength created in the direction of highest gain of the transmitting antenna is:

$$E = \frac{\sqrt{30G_T P_T}}{d} \text{ (volts/meter)} \tag{12.4}$$

The power picked up by a receiving antenna with a known capture area is:

$$P_R = A \frac{E^2}{240\pi} \text{ (watts)} \tag{12.5}$$

where A = effective capture area in square meters

If the gain of the receiving antenna is known instead of the capture area, the power picked up is:

$$P_R = \frac{G_R}{960} \left(\frac{\lambda E}{\pi}\right)^2 \text{ (watts)} \tag{12.6}$$

12.5 VERTICAL MONOPOLE

A variation on the dipole antenna is the vertical monopole (single-pole) antenna shown in Figure 12.9. Its overall length is half that of the dipole and, unlike the dipole, it is intentionally operated close to a large metal surface (called a *ground plane*), making it very convenient for automobile antennas.

The large conducting surface could be thought of as an electrical "mirror" to make the monopole think the other half of the dipole is still there. However, that is just one way to view the operation. The monopole has its own electrical and magnetic field patterns and its own set of impedances that vary with its length. These will depend somewhat on the size of the ground plane but even small surface areas with dimensions of a quarter wavelength are useful.

The radiation pattern of the monopole is shown in Figure 12.10. In three dimension, it would look something like a bagel sliced in half and placed face down on a plate. From an impedance point of view, the best length is about 0.25λ depending on element diameter. The input impedance at this length is a purely resistive 35 ohms. Notice that this is different from the 72 ohms of the dipole.

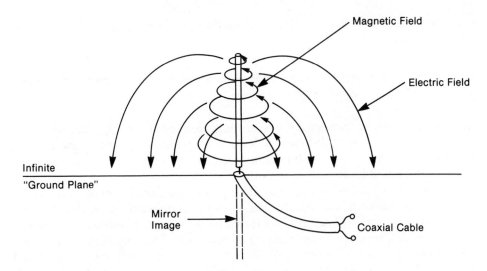

Figure 12.9 Vertical monopole uses a large conducting surface to replace the missing "mirror image" of the dipole.

The vertical antenna is very commonly used for the transmitting antennas of AM radio stations and very low frequency (VLF) navigation stations. The problem is that the soil in the middle of some field has to act as the ground plane and soil isn't the best conductor in the world. Therefore, the sites for these antennas must be carefully chosen by checking water and mineral content and soil type. Then long trenches must be dug and ground wires stretched radially out from the base of the antenna. The complete antenna "farm" can be very expensive but failure to do this will result in a large resistance in series

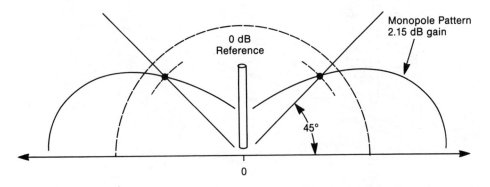

Figure 12.10 Radiation pattern of a vertical monopole with infinite ground plane.

with the radiation resistance that represents pure heating losses and wasted transmitter dollars.

12.6 ANTENNA ARRAYS

Antenna gain can be increased if two or more elements (either dipoles or monopoles) are used together. When this is done, a change in the radiation pattern always occurs and so we are still getting the gain by concentrating the radiation more in the one direction than the other.

Figure 12.11 shows two vertical monopoles separated by a half wavelength and fed in-phase with half of the signal power going to each. A receiver located some distance away will receive half its signal from each antenna but there will be a phase difference between the two components. The amount of phase shift will be proportional to the extra distance that the one signal has to travel and this, in turn, will depend on the azimuth angle of the receiver. If the receiver is directly in front of or behind the two monopoles, both distances are equal so the signals arrive in phase and add to each other to provide a very strong signal. If the receiver is out to the side and in line with the two elements, the one signal has to travel one-half wavelength further and so will arrive out of phase with the first. The result will be total cancellation. The resulting pattern is shown in Figure 12.12.

Higher gains and narrower beam widths can be obtained by adding more radiating elements and feeding a portion of the transmitter power to each. The resulting patterns are easily calculated with a simple computer program.

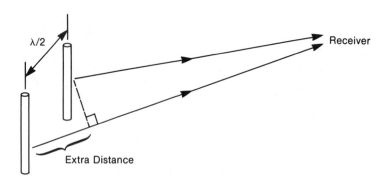

Figure 12.11 Two monopoles separated by a half wavelength and fed in-phase with half of the transmitter power.

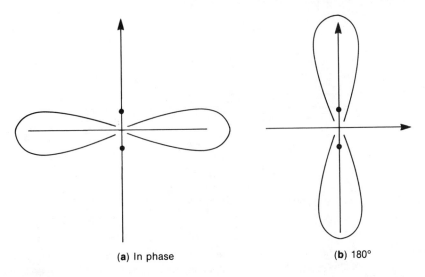

(a) In phase   **(b)** 180°

Figure 12.12   Radiation pattern looking down on the top of the two monopoles of Figure
12.11 is shown in (a). If the two elements had been fed power out of
phase, the pattern in (b) would be the result.

12.7 PASSIVE ELEMENTS

A second method of obtaining gain involves the use of passive elements placed
close to the driven element and cut to slightly different lengths. If these extra
elements are close to the driven element, they will be surrounded by its electric
and magnetic fields and so can pick up and reradiate the original signal. The
relative phase of this reradiated signal will depend on the spacing between the
driven and the passive elements and also on the overall length of the elements.
If the passive element is cut shorter than one-half wavelength, it will be capacitive. If cut longer, it will be inductive. Figure 12.13 shows a three-element antenna. The longer passive element acts as a reflector and the shorter one is a
director. The major radiation will be in the direction of the shorter element.
This antenna will have a gain of about 7 dB more than an isotropic radiator.
More director elements can be added to provide even more gain but there is a
penalty. As more elements are added, the antenna becomes much more frequency-sensitive. The extra gain is obtained but over an increasingly narrow
bandwidth.

   The presence of the extra metal close to the driven element changes its
impedance—it lowers it. To raise it back up to a value that will match a 300 ohm
balanced transmission line, a different driven element called a *folded dipole* is
often used. The radiation properties and gain of this element are exactly the
same as the ordinary dipole; only its impedance is different.

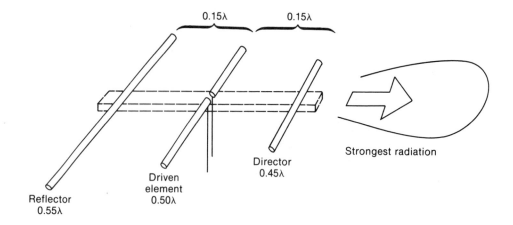

Figure 12.13 A passive reflector (longer) and director (shorter) can be placed close to a half-wave dipole to redirect some of the radiation and so increase the gain. This is known as a *Yagi antenna*, named after its developer.

It has already been stressed that any property of an antenna will be independent of whether it is used as a transmitting or a receiving device. If a Yagi antenna has a higher gain when transmitting, it must also show this when receiving. This means that the effective capture area of a Yagi must be larger than that for a dipole. Yet, when you stand in front of a Yagi antenna and look at it head on, the same as an incoming signal would, there is no visible difference. The explanation is shown in Figure 12.14. As the wave passes over the length of the antenna, it is progressively pulled in from greater distances. The passive elements pass their signals over to the only element that can consume the energy, through electric and magnetic fields—just like coupled tuned circuits of Chapter 3.

12.8 PARABOLIC REFLECTORS

For very high gains and narrow beamwidths, it becomes very tedious to connect large numbers of either passive or active elements. Besides, the bandwidth of these elaborate arrays becomes progressively smaller. Therefore, it becomes more attractive to use a simple radiator with a wide beamwidth and then use a curved metallic reflector to redirect its radiation.

The source or feed antenna must have some minimal directional properties. It should send the bulk of its signal in the general direction of the reflecting surface and not in the opposite direction; otherwise 50% of the potential gain

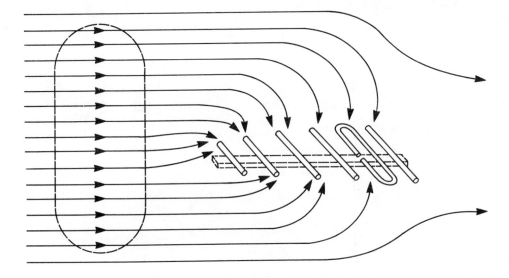

Figure 12.14 The capture area of a Yagi antenna is large because the wave is pulled in progressively along the full length of the structure.

will be lost. A dipole and single reflector are usually adequate. At microwave frequencies, a flared end (horn) on the waveguide is sufficient.

The reflector is a parabolic surface (parabolloid) and the feed antenna is positioned at its focal point. This combination satisfies the requirement that, after reflection, all of the wave must be in-phase over the surface of a plane that crosses the beam at right angles. This "phase-front" is shown in Figure 12.15.

Figure 12.15 can be a bit misleading. It might tend to suggest that the diameter of the dish plays no real part in the operation of the antenna. It might seem possible to use a smaller "dish" with a shorter focal length and move it closer to the feed antenna. After all, the ray paths in the diagram indicate that the full signal would still be properly reflected and the rays would all leave the reflector parallel to each other and in-phase.

The diameter does play a very important part. The gain and beamwidth of the overall antenna are totally dependent on the diameter. The reason is provided in the following statement and illustrated in Figure 12.16.

As the beam of an E/M signal travels through space it always tries to spread out. The reason is that every point on the front of the advancing wave acts as an independent, omnidirectional reradiator. The wave ahead is an interference pattern of all these sources. The wider the diameter of the advancing beam, the less it spreads out.

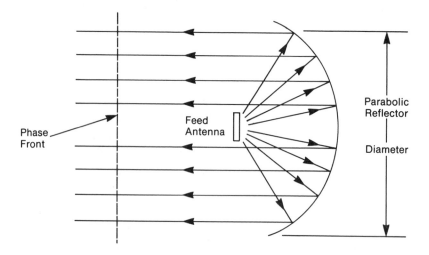

Figure 12.15 Parabolic reflector used to redirect the radiation from a simple "feed" antenna. High gains and narrow beamwidths are easily obtained by changing the diameter of the reflector.

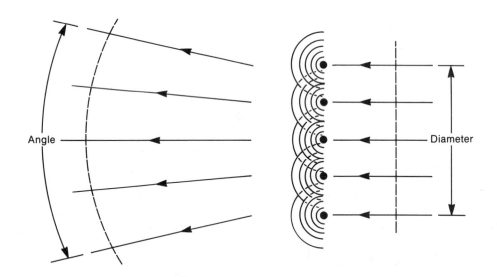

Figure 12.16 The divergence of any beam of E/M radiation is inversely proportional to the beam's diameter. Once the phase front becomes curved it maintains that angle.

The beamwidth of a large "dish" antenna is given by the following equation. This assumes that the feed antenna is able to spread its signal uniformly over the surface of the reflector, something that is very difficult to do without spilling over the edges. In practice, the beamwidth will be slightly wider than this.

$$\text{Beamwidth} = \frac{50\lambda}{\text{diameter}} \text{ (degrees)} \quad\quad (12.7)$$

When receiving, the effective capture area is slightly less than the dish's frontal area for the same reason. The feed does not collect signal from the full area uniformly.

A typical radiation pattern (in rectangular coordinates) for a parabolic antenna is shown in Figure 12.17. This shows the main radiation lobe and also a number of "sidelobes." These indicate that some of the radiation heads out in other directions—the result of some of the feed signal hitting the edges of the dish and reradiating, some missing the dish completely, and some reflecting off the parabolic surface and then hitting the feed antenna or its mechanical supports.

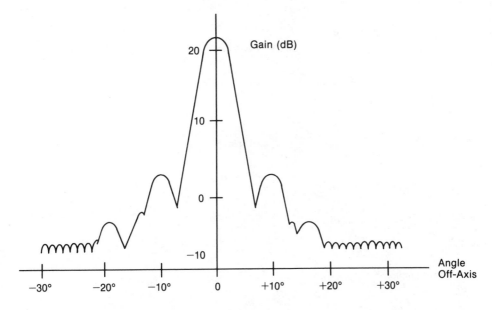

Figure 12.17 Radiation pattern of a dish antenna showing sidelobes. The graph coordinates are rectangular instead of polar to provide more detail of small angles.

12.9 PROPAGATION

Once the E/M signal leaves the immediate vicinity of the transmitting antenna, it is at the mercy of the media it travels in and any surrounding objects. In free space, the signal simply keeps spreading out. There is nothing to absorb the signal or make it bend or bounce. It just keeps getting weaker by 6.02 dB every time the distance doubles. A receiver always picks its signal up in a direct line from the transmitter with only one path involved.

Close to the surface of the Earth, life is considerably more difficult. The problems are illustrated in Figure 12.18.

Assuming a vertically polarized antenna at the surface of the Earth, the signal heads out in all directions (except straight up and down). Some of the signal heads straight out and listeners near by can pick up the signal directly—as long as there aren't any obstructions in the way. This part of the signal propagation is similar to the results in free space. There could be some extra attenuation due to actual heating losses in the moisture of the air but normally this is minimal. A listener at point B in the illustration could pick up the signal in this way.

As the radiation passes over the surface of the Earth, some will actually penetrate down into the soil. How deep will depend on the soil conditions and the frequency. The lower the frequency, the deeper the penetration—just like

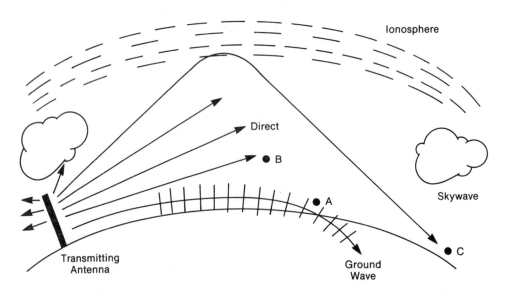

Figure 12.18 Radiation from a low-frequency antenna close to the Earth's surface could follow any of three general paths.

skin depth of a wire. Because of the higher dielectric constant of the soil, the signal travels slightly slower. The interaction of the signal in the soil and in the air above it causes the total signal to bend and follow the curvature of the Earth. How far this will continue depends on soil conditions, oceans, and the actual frequency itself. The navigation signals from 100 kHz LORAN-C transmitters will travel 2,000 or 3,000 miles. The 10 kHz OMEGA navigation signals cover the entire globe.

A third portion of the radiated signal misses the Earth's surface and heads up into the sky where it encounters the ionosphere—layers of charged particles located at various heights from 20 to 200 kilometers. What happens here depends again on the frequency. Above 30 MHz, the signal passes straight on through. At lower frequencies, there is an increasing chance that the change in velocity will cause some of the signal to slowly curve back down toward the Earth. What happens to any given frequency depends on the time of day, season of the year, the 11-year sunspot cycle, and the number of white hairs on the tail of a yak in Tibet.

Through this bending, signals can be received on Earth great distances away from the transmitter. The major problem is that the path's characteristics keep changing and several different paths to the same point could exist at once. Fading of the total signal or just some of the signal frequencies is the result. This adds to the excitement of shortwave radio reception.

This "sky wave" causes another problem. Long-range navigation signals rely on the "ground wave" for predictable position measurements. At night, the sky wave picks up in strength and interferes with the ground wave. A boat riding at anchor will often appear to be moving rapidly just at sunset if the navigation signals are to be believed.

There can still be some problems even at distances much closer to the transmitter. Nearby objects can obstruct or reflect the signal. The possibilities are shown in Figure 12.19. If an obstruction blocks the view between the transmitter and the receiver, some signal might still be received. Recalling the statement that every point on the front of an advancing wave acts as its own radiator, it seems reasonable that a signal can actually bend down behind an obstruction. The radio shadow will not be as long as a corresponding light shadow. Spend some time at the beach watching waves pass the end of a breakwall.

The opposite situation is also shown in this illustration. Even though a transmitting antenna can be seen directly, no radio signal might be received in some cases. The problem is that some of the antenna signal could be bouncing off the ground and then reaching the receiving antenna. The longer path taken by the bouncing signal results in a phase shift relative to the direct signal. The shift could be just enough to cancel the direct signal. Moving slightly forward or back, up or down will often clear up the problem. For mobile radios, however, this is a continual problem that limits the reception area in areas with many nearby reflecting surfaces.

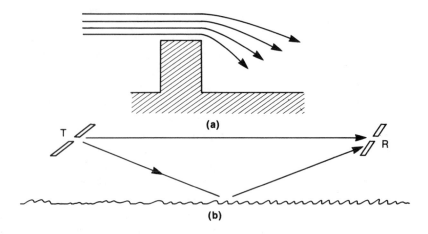

Figure 12.19 Radio waves can bend down behind an obstruction (a) to shorten the "shadow." But, in a direct line of sight, a bouncing wave (b) could cancel the direct wave.

12.10 REVIEW QUESTIONS

1. What special electrical properties does the metal for an antenna need?

2. If two CB operators exchanged their antennas, would there be any change in their transmitted and received signals?

3. If a dipole antenna is made very slighly longer than it should be, does it pick up any extra signal?

4. What does happen if you change the size of a dipole slightly?

5. What conditions must exist around an antenna before an electromagnetic wave can be sent out?

6. What is the approximate impedance of a short 75 MHz dipole antenna that is 1.6 meters end-to-end and uses 1.2 cm diameter tubing?

7. What is the effective capture area for a half-wave dipole antenna at 17.3 MHz?

8. Using the capture area calculated in question 7, find how much power would be received by a half-wave dipole located 65 km away from a second dipole radiating 10 watts at 17.3 MHz. Both antennas are oriented for best transmission and reception.

9. Repeat questions 7 and 8 but use a frequency of 34.6 MHz and then compare the received power levels at the two frequencies.

10. For a transistor to provide "gain," it must be connected to a DC power supply. An antenna can provide "gain" without a power supply. Why?

11. How much power would be picked up by a half-wave dipole at 108 MHz if it is in an electric field of 100 microvolts per meter?

12. Using Equation 12.3, calculate the power received at 4500 MHz by a receiving antenna with a 21 dB gain. The signal comes from a satellite radiating 1000 watts through a 25 dB gain antenna at a distance of 40,500 km.

13. Write a computer program to calculate the horizontal radiation pattern that would result if three vertical monopoles are fed equal power in-phase. The poles are spaced 0.4 wavelengths from each other in a straight line.

14. Calculate the beamwidth of a 4000 MHz satellite dish antenna that has a diameter of 3.5 meters.

15. What portion of an antenna's radiated energy is used for long-distance, short-wave listening? Why does the quality of short-wave signals change so much over periods of minutes, hours, months, and years?

SELECTED ANSWERS TO REVIEW QUESTIONS

Chapter 1

6. C-band, VHF, UHF, MF, HF, LF, L-band

9. 2.26 μV

10. 1.43 dB

12. 1.2, 4.5, 6.3, 9.8, 2.4, 9.0, 12.6, 19.6, 5.7, 7.5, 11.0, 10.8, 14.3, 16.1, 3.3, 5.1, 8.6, 1.8, 5.3, 3.5 kilohertz

14. 0.3 meters

15. a. 209.8 meters

    b. 15.63 meters

    c. 2.80 cm

16. 0.177 microsecond

Chapter 2

8. 138.4 Ω

9. 0.0542 mm

10. 246.6 Ω

11. Approximate inductance = 0.8 μH; loss = 0.58 dB

13. 0.0367 μF

15. 22.64 $-$ j102.6 ohms

Chapter 3

3. 24 dB/octave

6. 19.43 dB, 83.9°, 37.58 nano sec.

9. 250 Ω to keep input and output $Q = 2.5$

11a. 314 Ω in parallel with $+j\ 92.35$ Ω

12a. $238.7 + j\ 426$ Ω

14. 40.8 Ω

Chapter 4

5. 1425 Ω

6. 32.83 kHz

10. $C_1 = 3854$ pF, $C_2 = 481.7$ pF, $L = 1.700\mu$H

11. Ratio $= 2.622$

Chapter 5

1. LSB $=$ 824.8 kHz; USB $=$ 835.2 kHz; BW $=$ 10.4 kHz

3. 1.3kHz   10 volts
   2.6 kHz   4.0 volts
   23.7 kHz   20 volts
   25.0 kHz   25 volts
   50.0 kHz   25.0 volts
   26.3 kHz   20 volts
   $m = 1.6$

8. 16.67% of total power

10. a.   3.5 watt carrier

    b.   0.315 watt each sideband

    c.   4.13 watt total power

Chapter 6

4. $m = 6.25$

7. $C = 7500$ pF

9. Bandwidth $= 58$ kHz; $m = 4.8$

10. 14 dB quieter

Chapter 7

  7. 21 volts p-p

  8. RF power = 0.648 watt; DC current = 87.1 mA

 10. 40 dB attenuation (30.86 dB minimum)

Chapter 8

  6. Lowest Frequency = 562.7 kHz
      BW   = 39.79 kHz
      Highest Frequency = 1453 kHz
      BW   = 265.3 kHz

  9. Image Frequency   = 3.10 MHz
      Intermediate Frequency = 1.05 MHz

Chapter 9

  3. Gain = 2.143 MHz/volt

  5. Voltage = 2.30 volts

  7. Frequency   = 1.2kHz
      DC component = 1.8 volts
      AC component = 0.225 volts p-p

 11. Reference Frequency = 0.2 MHz
      Dividers set to 20, 21, 22 23, 24, 25

 12. Frequency = 43.5 MHz

Chapter 10

  1. Velocity = 179.3×10^6 meters/second

  3. Z_o = 122.1 Ω

  5. Z_o = 218.9 Ω

  7. a. Reflection Coefficient = 0.0625

     b. Line VSWR = 1.133

     c. Highest Voltage = 37.19 volts
        Lowest Voltage = 32.81 volts

  d. Highest Impedance = 85 Ω
    Lowest Impedance = 66.2 Ω

10. Z_o = 141.4 Ω, length = 19.74 meters

11b. Input Impedance = 41 + j16.3 Ω

14. Loss approximately 1.5 dB

16. Guide wavelength = 14.4 cm

Chapter 11

 5. Bandwidth = 250 Hz

 6. Bandwidth = 250 Hz

15. Maximum Data Rate = 66,580 bps

Chapter 12

 6. Input impedance approximately $50 - j125$ Ω

 7. Capture Area = 39.24 sq. meters[6]

 8. Received Power = 1.212 $\times$ 10^{-8} watt

11. Received Power = 13.36 $\times$ 10^{-12} watt

14. Beamwidth = 1.07°

INDEX